T0211967

Basic Simulation Models
of Phase Tracking Devices
Using MATLAB

Synthesis Lectures on Communications

Editor
William Tranter, *Virginia Tech*

Basic Simulation Models of Phase Tracking Devices Using MATLAB
William Tranter, Ratchaneekorn Thamvichai, and Tamal Bose
2010

Joint Source Channel Coding Using Arithmetic Codes
Dongsheng Bi, Michael W. Hoffman, Khalid Sayood
2009

Fundamentals of Spread Spectrum Modulation
Rodger E. Ziemer
2007

Code Division Multiple Access(CDMA)
R. Michael Buehrer
2006

Game Theory for Wireless Engineers
Allen B. MacKenzie, Luiz A. DaSilva
2006

Basic Simulation Models of Phase Tracking Devices Using MATLAB

William Tranter, Ratchaneekorn Thamvichai, and Tamal Bose

ISBN: 978-3-031-00548-0   paperback
ISBN: 978-3-031-01676-9   ebook

DOI 10.1007/978-3-031-01676-9

A Publication in the Springer series
SYNTHESIS LECTURES ON COMMUNICATIONS

Lecture #5
Series Editor: William Tranter, *Virginia Tech*
Series ISSN
Synthesis Lectures on Communications
Print 1932-1244   Electronic 1932-1708

Basic Simulation Models of Phase Tracking Devices Using MATLAB

William Tranter
Virginia Tech

Ratchaneekorn Thamvichai
Saint Cloud State University

Tamal Bose
Virginia Tech

SYNTHESIS LECTURES ON COMMUNICATIONS #5

ABSTRACT

The Phase-Locked Loop (PLL), and many of the devices used for frequency and phase tracking, carrier and symbol synchronization, demodulation, and frequency synthesis, are fundamental building blocks in today's complex communications systems. It is therefore essential for both students and practicing communications engineers interested in the design and implementation of modern communication systems to understand and have insight into the behavior of these important and ubiquitous devices. Since the PLL behaves as a nonlinear device (at least during acquisition), computer simulation can be used to great advantage in gaining insight into the behavior of the PLL and the devices derived from the PLL.

The purpose of this Synthesis Lecture is to provide basic theoretical analyses of the PLL and devices derived from the PLL and simulation models suitable for supplementing undergraduate and graduate courses in communications. The Synthesis Lecture is also suitable for self study by practicing engineers. A significant component of this book is a set of basic MATLAB-based simulations that illustrate the operating characteristics of PLL-based devices and enable the reader to investigate the impact of varying system parameters. Rather than providing a comprehensive treatment of the underlying theory of phase-locked loops, theoretical analyses are provided in sufficient detail in order to explain how simulations are developed. The references point to sources currently available that treat this subject in considerable technical depth and are suitable for additional study.

KEYWORDS

PLL, PLL simulation, frequency tracking, phase tracking, signal acquisition, signal tracking, synchronization, digital demodulation, Costas PLL

Contents

Preface

The Phase-Locked Loop (PLL), and many of the devices used for frequency and phase tracking, carrier and symbol synchronization, demodulation, and frequency synthesis, are fundamental building blocks in today's complex communications systems. It is therefore essential for both students and practicing communications engineers interested in the design and implementation of modern communication systems to understand and have insight into the operating characteristics of these important and ubiquitous devices. Since the PLL behaves as a nonlinear device, at least during acquisition, a traditional pencil and paper analysis of the PLL operating in the acquisition mode are extremely difficult. However, many important acquisition characteristics of the PLL, such as acquisition time, must be determined. Computer simulation can be used to great advantage in these situations to gain insight into the behavior of the PLL and the devices derived from the PLL.

The purpose of this Synthesis Lecture is to provide basic theoretical analyses of PLL and devices derived from the PLL and their simulation models suitable for supplementing undergraduate and graduate courses in communications and for self study by practicing engineers. A significant component of this book is a set of basic MATLAB-based simulations that illustrate the operating characteristics of these devices and enable the reader to investigate the impact of varying system parameters. This Synthesis Lecture by no means provides a comprehensive treatment of the underlying theory of phase-locked loops, and we make no attempt to be rigorous or complete. There are many excellent books currently available that treat this subject in considerable technical depth. In this synthesis lecture, however, theoretical analyses are provided in sufficient detail in order to explain how simulations are developed.

Following the introductory chapter, the theory of the basic PLL operating in the absence of noise is discussed in Chapter 2 to the point that MATLAB simulation models can be developed and understood. Both tracking and acquisition operating modes are considered. First-order, second-order and third-order PLLs are presented. Because of the importance of the second-order PLL, a number of variants of the basic second-order structure are presented. These include the perfect second-order PLL, the imperfect second-order PLL, and the second-order PLL with transport delay.

Chapter 3 illustrates more advanced structures derived from the simple PLL. The Costas PLL is first explored. An important application of the Costas PLL is the demodulation of BPSK. This is followed by a discussion of demodulation structures for QPSK. The last topic treated in this chapter is the generalization to an N-Phase tracking loop.

Chapter 4 and 5 presents the simulation of the structures presented in Chapters 2 and 3. The underlying mathematics, and the approximations involved, is the subject of Chapter 4. The actual MATLAB simulations are developed in Chapter 5. The simulation methodology based on separation of a simulation program into preprocessors, simulators and postprocessors are included for all simulation examples.

The operation and performance of the PLL in the presence of noise is the subject of Chapter 6. The linear analysis of the PLL with input additive noise is performed. The noise equivalent bandwidth and the loop Signal-to-Noise Ratio (SNR) are discussed for the PLLs considered in Chapter 2.

A number of Appendices are included. Appendix A presents a brief overview of complex envelopes and the use of complex envelopes for representing bandpass signals. This appendix also contains a brief discussion of phase detector models and implementations of these models. Appendix B treats the approximation of continuous-time integration using discrete-time techniques. The focus is on trapezoidal integration. Appendix B also contains a discussion on the MATLAB implementation of loop filters including the loop filters for the perfect second-order loop, the imperfect second-order loop, and the perfect third-order loop. The development of the simulation code for the baseband PLL model is the focus of Appendix D. Appendix C presents a few of the MATLAB examples recast as SIMULINK simulations in order to provide the student with an introduction to the use of SIMULINK in the context of PLL simulations. Appendix D provides a listing of the m-files necessary to run all of the simulations contained in this book. The complete m-files can be downloaded from the Morgan and Claypool website, which has the URL

http://www.morganclaypool.com/page/pll

A few words are in order concerning the origin of this lecture. A number of years ago one of the authors (William Tranter) of this lecture was, with the help of several colleagues, developing a textbook on the simulation of communications systems. Because of the importance of the PLL in the implementation of modern communication systems, a chapter in this book dealing with PLLs was considered necessary. In addition, this chapter also served as an introduction to the simulation of nonlinear systems. This book was later published by Prentice Hall and much of this synthesis lecture is based on parts of this book. (This is especially true of Chapter 2, Appendix A, and Appendix B.) We appreciate Prentice Hall giving us permission to use this material. For students having an interest in the broader subject of simulation systems should consult the following:

William H. Tranter, K. Sam Shanmugan, Theodore S. Rappaport, and Kurt L. Kosbar, Principles of Communication Systems Simulation with Wireless Applications, Prentice Hall, Professional Technical Reference, 2004, IBSN: 0-13-494790-8.

This book, like this synthesis lecture, is MATLAB based.

A word concerning the history of this synthesis lecture is perhaps in order. At about the same time the simulation book was being developed, Bill Tranter had the good fortune to spend six months at the University of Canterbury as an Erskine Fellow in Christchurch, New Zealand. While at Canterbury, he taught a course in communications system simulation. Several students in this class, especially a Ph.D. student, Ms. Katharine Holdsworth, asked a number of excellent questions concerning the operational characteristics of a number of devices derived from the second-order PLL. The students desired to gain insight into loop operations and mentioned that it would be extremely useful to see waveforms present at various points within the system and to observe the effect of varying loop parameters. To answer these "what if" questions is a principle motivation for simulation, and we began a process of developing simulation models of various structures based on the PLL. In addition, Professor Desmond Taylor, a friend and colleague at Canterbury University, indicated that in many academic environments, it is simply not possible for students to take courses covering all of the necessary subjects in which communications engineers need an introductory understanding. We discussed the subject of phase-locked devices and agreed that this was one of those subjects. It was out of these discussions that this synthesis lecture was born.

In addition to Katharine Holdsworth and Desmond Taylor, we wish to thank the co-authors of the Prentice Hall simulation book (K. Sam Shanmugan, Theordore S. Rappaport, and Kurt L. Kosbar) for all of the knowledge gained while working with them. We also thank Bernard Goodwin of Pearson Publishing for giving us permission to use material from the simulation book in this lecture.

To conclude, a number of people have asked why MATLAB, rather than SIMULINK, was the tool of choice in the development of the simulations contained in this synthesis lecture. There are a variety of answers to this question. First, MATLAB is widely used in many engineering educational programs. Since MATLAB is a very high level general-purpose programming language, it can be used in a wide variety of course settings. It is both powerful and computationally efficient. The basic language, together with the many toolboxes that have been developed by both The Math Works and a number of third-party vendors, provides a rich library of routines available for systems analysis and for program development. The conciseness of MATLAB code makes it possible to express complex ideas using a very few lines of code. An important purpose of many simulations, including those presented here, is to allow one to view the waveforms present at various points in a system and to view operational characteristics derived from these waveforms. MATLAB's extensive library of graphics routines makes it possible to easily develop simulation postprocessors for viewing the graphical presentations generated by a simulation. While it would be possible to develop the simulations contained here in almost any computer language, the use of MATLAB makes it somewhat more practical for course applications. Suffice it to say that, if MATLAB were not available, this book would not exist.

MATLAB, because of the conciseness of the code, is also a very nice prototyping language. One often develops a signal processing algorithm, which might be the simulation code for a subsystem, quickly and easily in MATLAB. After testing the algorithm to ensure correctness, it can then be easily converted to another language, such as C or C++, if desired. In addition, we often develop C-language simulation programs that use MATLAB-based postprocessors. By taking advantage of MATLAB's data handling and graphical capabilities, considerable time is saved. MATLAB-based preprocessors are also practical for use with simulations written in another language.

Blacksburg, VA
August 2010

William H. Tranter
Ratchaneekorn Thamvichai
Tamal Bose

CHAPTER 1

Introduction

The phase-locked loop (PLL), and many synchronization and demodulation devices that can be derived from the basic PLL, are almost ubiquitous in today's complex communications systems. The basic PLL structure is a maximum-likelihood estimator of the instantaneous phase deviation of an input bandpass signal in the presence of noise (16). The PLL is a device that attempts to synchronize the phase and frequency of a controllable oscillator with those of the input signal. In its most basic form, the PLL is an effective analog FM demodulator. The Costas PLL, which is a simple extension of the basic PLL, can be used for the demodulation of analog double-sideband suppressed carrier signals. In addition, the Costas PLL can be used for the demodulation of a digital signal in which binary phase-shift keying (BPSK) is the modulation format. A simple modification of the Costas PLL results in a demodulator for quadrature phase-shift keying (QPSK). The family of devices grows as other demodulation schemes are considered. In addition, PLLs form the basis of many of the carrier and bit synchronization sub-systems that are important components in the implementation of digital communications systems. It is therefore appropriate for students of communication theory to understand the operating characteristics of these important system building blocks.

The PLL is inherently a nonlinear system and is therefore difficult for beginning students of communications theory to fully comprehend and appreciate. A simple linear model may be used for PLL analysis for the case in which the PLL is *in lock*, but analysis of the acquisition behavior cannot be accomplished using the simple linear model and nonlinear analysis techniques are necessary. Such analysis techniques are well beyond the scope of most undergraduate courses in communication theory. It is possible, however, to use simulation in order to gain significant insight into PLL operation. In addition, simulation-based analysis can be used to solve important problems such as determining the time required to achieve phase lock when noise or interference is present. The following material is intended to give students and practicing engineers in the communications area a set of very basic PLL models that can be used in a simulation-based laboratory environment. Experiments can be performed by placing various signals on the loop input and observing the waveforms present within the loop that result from these signals. Loop parameters are easily changed and the results are observed. Therefore, parametric analyses and design studies can quickly be conducted.

Simulation preprocessors, simulation programs and simulation postprocessors are given for each PLL model discussed in the following chapters. All of these are developed in MATLAB. The preprocessors allow the simulation user to easily establish the system parameters and the parameters necessary for managing the simulation. The simulation program executes the simulation. The simulation postprocessor allows the user to easily generate simulation results such as time-domain waveforms and phase-plane plots. The simulation postprocessors included here are menu-driven and can easily be modified and expanded in order to allow one to examine any of the waveforms present in the loop structure. One can, of course, simply issue the appropriate MATLAB graphics command to display the desired waveform thereby bypassing the postprocessors provided.

This book is designed to meet a number of needs. It can be used as a self study supplement to basic communications courses by those students who desire to gain familiarity with the operating characteristics of various phase-locked systems. The MATLAB PLL models included in later chapters also provide the resources for in-class demonstrations. In addition, this book can be used in a variety of other ways. For those designing synchronizers or demodulators based on the PLL, the models included here will hopefully serve as a starting point. For one engaged in the study of communication theory, the basic models included in this book can be used as the building blocks of more complex systems.

The simulations presented here serve as examples of simple simulations and will hopefully help students face with the problem of developing their own simulations of communications systems. All MATLAB programs can be found at `http://www.morganclaypool.com/page/pll`.

A word is in order concerning the choice of MATLAB as a simulation tool. MATLAB, in the opinion of the authors, has become the computational tool of choice in engineering curricula. It is powerful and has a rich library of basic routines that satisfy the computational requirements for most engineering applications. The many toolboxes available provide the resources necessary for more specialized applications. An affordable student edition places the capabilities of MATLAB within the resources of individual undergraduate students. In addition, two attributes of MATLAB make it ideal for the application at hand; the high-level syntax of MATLAB makes it possible to express complex ideas very concisely and the extensive library of graphics commands allows for easy generation of a wide range of graphical output. Postprocessors, which by their very nature require extensive graphical resources, are therefore easily developed using MATLAB. As a matter of fact, we have, in a number of projects, used MATLAB-based postprocessors in simulations with simulation programs written in C, C++, or even FORTRAN.

1.1   OUTLINE OF THE BOOK

Chapter 2 introduces the theory of the basic PLL operating in the absence of noise. Since a large number of books have been developed which treat the subject of PLLs in considerable detail, no effort to be either rigorous or complete has been made here. Instead, an attempt has been made to simply develop the theory to the point that MATLAB models can be developed and understood. Both tracking and acquisition modes of PLL operation are analyzed. First-order, second-order and third-order PLLs are considered. Because of the importance of the second-order PLL, a number of variants of the basic second-order structure are presented. These include the perfect second-order PLL, the imperfect second-order PLL, and the second-order PLL with transport delay. In addition, several different phase detector models (sine-wave, triangular-wave and sawtooth-wave) are included in the second-order PLL simulation models so that the student can see the effect of using phase detectors having different characteristics. The student can also see different modeling approaches.

Chapter 3 illustrates more advanced structures based on the simple PLL. The Costas PLL is first explored. The application is the demodulation of BPSK. This is followed by a discussion of structures for QPSK loop. The last topic treated in this chapter is an N-Phase tracking loop.

Chapters 4 and 5 fulfill the purpose of this book by providing MATLAB simulation examples for each of the PLL architectures considered in Chapters 2 and 3. The simulation models are presented in Chapter 4 and examples based on these simulation models are contained in Chapter 5. Preprocessors, simulators and postprocessors are included for each architecture. Examples of the output generated by each example is given.

Chapter 6 studies the operation and performance of the PLL in the presence of noise. The linear analysis of PLL with input additive noise is considered. The noise equivalent bandwidth and the loop Signal-to-Noise Ratio (SNR) are discussed for PLLs considered in Chapter 2. The non-linear analysis of PLL with additive noise and the analysis of PLL under the presence of phase noise are briefly discussed.

A number of Appendices are included. Appendix A presents a brief overview of complex envelopes and the use of complex envelopes for representing bandpass signals. Appendix A also contains a description of phase detector models and illustrates the application of complex envelope techniques to the development of phase detector models. Appendix B treats the approximation of continuous-time integration using discrete-time techniques. Trapezoidal integration is exclusively used in the MATLAB models developed in the earlier chapters. Also included in Appendix B is the development of the simulation code for both the loop filters and for the baseband VCO model. Appendix C presents a few of the MATLAB examples recast as SIMULINK simulations. Appendix D lists the set of m-files necessary to run the simulations contained in this book. All MATLAB and SIMULINK simulations can be found at `http://www.morganclaypool.com/page/pll`.

1.2   A WORD OF WARNING

Simulation, properly used, is an effective technique for gaining insight into the operating characteristics of complex communications systems, for performing parametric studies and for evaluating design alternatives. A properly developed simulation also serves, at least in some ways, as a laboratory for playing the many "what if" games necessary to satisfy our intellectual curiosity concerning how systems perform in a variety of operating environments. The outcome of these "what if" questions often suggest new analysis techniques or designs. These are only a few of the worthwhile benefits derived from the use of simulation. One should be warned that simulation can, unfortunately, lead to a false sense of security. After running a number of simulations and observing the results, one often incorrectly jumps to the conclusion that they fully understand the system being simulated. While this may be true in some cases, it may also not be true. Each simulation is in reality a special case analysis based upon the assumptions that define the system architecture and the signal and system parameters used for that particular simulation model. One must be careful not to extrapolate these results to the point of making general conclusions about how the system will operate beyond the set of parameters actually simulated. The student must always keep in mind that just as a physical system is often simplified in the process of deriving an analytical model, additional approximations and simplifications are typically used in moving from an analytical model to a simulation model.

A student desiring a deep understanding of any subject, phase-locked devices included, must spend some effort studying the analysis techniques used in the subject under study. A good understanding of phase-locked devices demands that the student be familiar with some of the basic literature that defines the subject. The references listed throughout Chapter 2 are a starting point in PLL theory. A study of this material allows analysis skills to be developed. Simulation is most powerful when used hand-in-hand with analysis.

In addition, when one uses simulation for design and analysis studies, one must be careful not to base important decisions on simulation results until those simulation results have been validated. Validation is the process of insuring that the simulation results are reasonable and consistent with known theory. Validation therefore requires knowledge of analysis methods and familiarity with the literature. In summary, simulation is a valuable tool in the analysis and design process, but it should never viewed as a replacement for analytical skill.

1.3   ORIGINS OF THIS SYNTHESIS LECTURE AND A REFERENCE

As discussed in the Preface, this synthesis lecture was developed in parallel with the simulation book,

William H. Tranter, K. Sam Shanmugan, Theodore S. Rappaport, and Kurt L. Kosbar, *Principles of Communication Systems Simulation with Wireless Applications*, Prentice Hall, Professional Technical Reference, 2004, ISBN: 0-13-494790-8.

This book covers the basic theory of many of the techniques used in this synthesis lecture and parts of this synthesis lecture follow closely material contained in this book. In addition, since this book covers in depth the theory and general methodology of time-driven simulation, it contains much of the background material contained in this synthesis lecture. It may serve as a useful reference for students wishing more detailed material on simulation methodologies. The authors thank Pearson Publishing for allowing us to use this material in this synthesis lecture.

CHAPTER 2

Basic PLL Theory

In this chapter[1], we first consider the basic theory of the PLL and explore some of its operating characteristics. No attempt is made to present a complete or unified treatment since this is more than adequately accomplished in the literature. There are a number of excellent textbooks that treat the PLL in considerable detail as listed in Bibliography. Our aim, however, is to develop the theory to a point that allows us to develop and understand a number of simple models for these interesting and useful devices. We then turn our attention in the following chapter to several important structures for demodulation and tracking that are derived from the basic PLL. These include the Costas PLL, the QPSK tracking loop, and the N-phase tracking loop. MATLAB simulations based on these models are then developed as described in Chapter 4 and 5.

2.1   BASIC PHASE-LOCK LOOP CONCEPTS

The Phase-Lock Loop consists of following three basic components as illustrated in Figure 2.1:

1. Phase detector: The phase detector compares an input signal with an output from a voltage-controlled oscillator. Its output is proportional to the phase difference between these two signals.

2. Loop filter: The loop filter controls the loop characteristics and, in general, smooths the output from the phase detector and applies it to the voltage-controlled oscillator. As will be seen in the subsequent sections, the choice of the loop filter has an impact on the properties of the loop and the performance of the PLL.

3. Voltage-Controlled Oscillator (VCO): The VCO is an oscillator in which the output frequency deviation is proportional to the input signal level.

The PLL operates in one of two modes; acquisition and tracking. In the *acquisition* mode, the PLL attempts to synchronize both the frequency and phase of the VCO output with those of an input signal. The phase errors between these two signals can be quite large. In this situation, the PLL is a nonlinear system and nonlinear analysis techniques are necessary for analysis of the acquisition behavior. Nonlinear analysis techniques are difficult and well beyond the scope of most undergraduate courses. It is possible, however, to use computer simulation in order to gain significant

---

[1]Much of the material in this chapter is based on Chapter 6 of

William H. Tranter, K. Sam Shanmugan, Theodore S. Rappaport, and Kurt L. Kosbar, *Principles of Communication Systems Simulation with Wireless Applications*, Prentice Hall, Professional Technical Reference, 2004, ISBN: 0-13-494790-8.

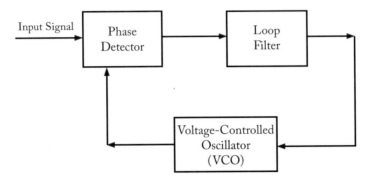

Figure 2.1: Basic PLL Block Diagram

insight into the PLL operation. In the *tracking* mode, the phase errors are often small and analysis using a simple linear model can be used to provide acceptable and useful results. The standard loop parameters, as we will see, are defined in terms of the linear model.

2.1.1   BASIC PLL MODEL

The basic model of a PLL is illustrated in Figure 2.2. The input signal is assumed to be

$$x_{in}(t) = A_c \cos[2\pi f_c t + \phi(t)]. \tag{2.1}$$

The signal at the output of the VCO is assumed to have the form

$$x_{vco}(t) = -A_v \sin[2\pi f_c t + \theta(t)]. \tag{2.2}$$

The characteristics of the phase detector determine, in part, the operating characteristics of the PLL. The choice of the phase detector to be used in a given situation depends on the application. The most common phase detector, a sinusoidal phase detector, is one whose output is proportional to the sine of the phase error. The sinusoidal phase detector consists of a multiplier and a lowpass filter with a gain of 2 as shown in Figure 2.3. The functions of the lowpass filter are to remove the second harmonic of the carrier frequency and the factor of 1/2 that results from the multiplication $\cos(x)\sin(y)$.

Assuming a sinusoidal phase detector, the output of the phase detector is

$$e_d(t) = A_c A_v \sin[\phi(t) - \theta(t)] \tag{2.3}$$

where the quantity $\phi(t) - \theta(t)$ is referred to as *phase error*. Later, we will denote the phase error by $\psi(t)$, but for now, it is better to keep all of our expressions in terms of input phase, $\phi(t)$, and VCO phase, $\theta(t)$. We desire the VCO phase to be an *estimate* of the input phase, and, therefore, a proper operation of the PLL is such that the phase error is driven towards zero. We will see later

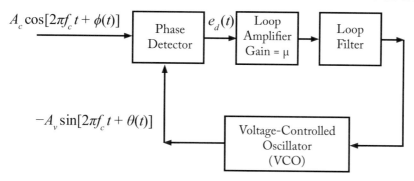

Figure 2.2: Basic Phase-Locked Loop Model

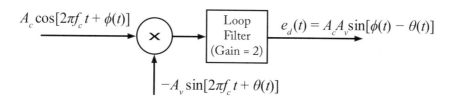

Figure 2.3: Phase Detector Model

how this is accomplished. The steady-state phase error may or may not be zero depending upon the characteristics of the input signal and the loop filter. The dependency of the steady-state phase error on the relationship between the input phase deviation and the parameters of the PLL will be explored in a later section of this chapter.

Note that the PLL input and the VCO output are in phase quadrature (90° out of phase) for $\phi(t) = \theta(t)$. This is required if the phase detector output is to be an *odd function* of the phase error, $\sin[\phi(t) - \theta(t)]$. It is easily seen that if the cosine functions are used in both (2.1) and (2.2), with the arguments unchanged, the phase detector output would be proportional to $\cos[\phi(t) - \theta(t)]$, which is an even function of the phase error. This would, of course, yield a system in which negative phase errors are not distinguishable from positive phase errors.

After multiplication by a loop gain, μ, the phase detector output is filtered by a loop filter having the transfer function $F(s)$ and a unit impulse response $f(t)$. The input to the VCO is therefore given by

$$e_{vco}(t) = \int_{-\infty}^{\infty} \mu e_d(\lambda) f(t - \lambda) d\lambda = \int_{-\infty}^{\infty} \mu A_c A_v \sin[\phi(\lambda) - \theta(\lambda)] f(t - \lambda) d\lambda \qquad (2.4)$$

which is simply the convolution of the loop filter impulse response with the loop filter input. The output of the VCO is (2.2) with the frequency deviation proportional to the VCO input signal. In other words, the VCO phase deviation is

$$\frac{d\theta}{dt} = 2\pi K_d e_{vco}(t) \tag{2.5}$$

where K_d is known as the VCO constant and has units of Hertz per volt. Solving (2.5) for $\theta(t)$ and substituting (2.4) for the VCO input yields

$$\theta(t) = 2\pi K_d \mu A_c A_v \int_{-\infty}^{t} \int_{-\infty}^{\infty} \sin[\phi(\lambda) - \theta(\lambda)] f(\tau - \lambda) d\lambda d\tau \tag{2.6}$$

This result is the nonlinear integral equation relating the phase deviation of the input $\phi(t)$ to the VCO phase deviation $\theta(t)$. Keep in mind that the impulse response of the loop filter is still arbitrary. Before proceeding, we make a small simplification in (2.6). The loop gain, G, is defined as the product of the constants in (2.6). Thus,

$$G = 2\pi K_d \mu A_c A_v \tag{2.7}$$

This gives

$$\theta(t) = G \int_{-\infty}^{t} \int_{-\infty}^{\infty} \sin[\phi(\lambda) - \theta(\lambda)] f(\tau - \lambda) d\lambda d\tau \tag{2.8}$$

as the final form of the general loop equation.

2.1.2   NONLINEAR PLL PHASE MODEL

It is apparent from (2.8) that the relationship between $\theta(t)$ and $\phi(t)$ does not depend in any way upon the carrier frequency f_c, and, therefore, the carrier frequency need not be considered in the analysis. We therefore seek a model that establishes the proper relationship between $\theta(t)$ and $\phi(t)$, without consideration of the carrier frequency. This model, shown in Figure 2.4, is known as the *nonlinear phase model* of the PLL. It is a nonlinear model because of the sinusoidal nonlinearity. It is a phase model because the model establishes the relationship between the input phase deviation and the VCO phase deviation rather than establishing the relationship between the actual loop input and VCO output signals as expressed by (2.1) and (2.2), respectively.

It is important to remember that the input to the model illustrated in Figure 2.2 is the actual bandpass signal present in the system under study while the input to the nonlinear phase model shown in Figure 2.4 is the phase deviation of the input bandpass signal. Of course, if the phase deviation, $\phi(t)$, and the carrier frequency are known, (2.1) is completely determined. Similarly, if the VCO phase deviation, $\theta(t)$, and the carrier frequency are known, the VCO output (2.2) is completely determined. Thus, the nonlinear phase model expresses the important quantities of interest. In simulation applications, the nonlinear phase model pays an additional dividend. Since the loop input and VCO output phase deviations are lowpass quantities, they can be sampled at a much lower sampling rate than the signals expressed by (2.1) and (2.2), which are bandpass signals.

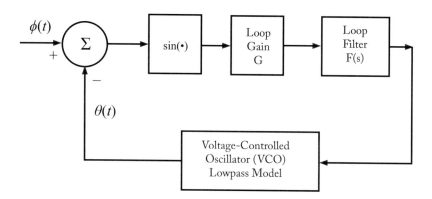

Figure 2.4: Nonlinear Phase Model of the Phase-Locked Loop

At this point, another word is in order concerning the lowpass filter used to remove the second harmonic of the carrier produced by the multiplier in the phase detector model. This filter is simply part of a conceptual model and need not be present in the physical device. It is easily seen that this filter may be eliminated. Equations (2.5) and (2.6) show that the VCO operates as an integrator. Since an integrator is a lowpass filter, which has an infinite gain at $f = 0$ and a unity gain at $f = 1/2\pi$ Hertz, the VCO will prevent the second harmonic of the carrier frequency from propagating around the loop and appearing at the VCO output. In the absence of a lowpass filter in the phase detector, the loop phase error must be observed by placing a lowpass filter external to the loop as shown in Figure 2.5. The purpose of the filter is to remove the troublesome second harmonic so that the phase error is observed at the filter output. Placing the lowpass filter outside of the loop will have the added benefit of removing the influence of this filter on loop dynamics.

In the preceding development, we use f_c for both the input carrier frequency and the quiescent VCO frequency[2]. In other words, we assumed that these two frequencies were equal. If the VCO is detuned from the input carrier frequency by an amount f_Δ, we can then represent the VCO output as

$$x_{vco}(t) = A_v\sin[2\pi f_c t + 2\pi f_\Delta t + \theta(t)] = A_v\sin[2\pi f_c t + \theta_1(t)] \tag{2.9}$$

so that the frequency offset is absorbed into the phase deviation as a ramp in phase with slope $2\pi f_\Delta$. In the work to follow, we assume $f_\Delta = 0$ without loss of generality.

---

[2]The quiescent frequency of an oscillator is the frequency of oscillation with zero input excitation.

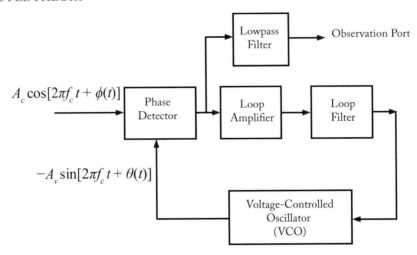

Figure 2.5: System for Observing the Phase Error in the Bandpass Model

2.2   PLL LINEAR PHASE MODEL

If the phase error is small so that the linear approximation

$$\sin[\phi(t) - \theta(t)] \approx \phi(t) - \theta(t) \qquad (2.10)$$

can be made, the loop equation, (2.8), becomes

$$\theta(t) = G \int_{-\infty}^{t} \int_{-\infty}^{\infty} [\phi(\lambda) - \theta(\lambda)] f(\tau - \lambda) d\lambda d\tau \qquad (2.11)$$

This results in the linear phase model of the PLL shown in Figure 2.6.

Taking the Laplace transform of (2.11), recognizing that integration is equivalent to division by s and that convolution in the time domain is equivalent to multiplication in the frequency (or Laplace) domain, yields

$$\Theta(s) = G[\Phi(s) - \Theta(s)]\frac{F(s)}{s} \qquad (2.12)$$

The transfer function, $H(s)$, relating the VCO phase to the input phase is therefore given by

$$H(s) = \frac{\Theta(s)}{\Phi(s)} = \frac{G\frac{F(s)}{s}}{1 + G\frac{F(s)}{s}} \qquad (2.13)$$

or

$$H(s) = \frac{GF(s)}{s + GF(s)} \qquad (2.14)$$

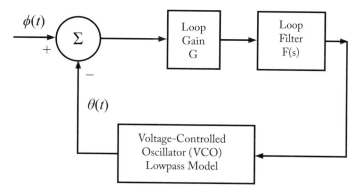

Figure 2.6: Linear PLL Phase Model

This is also called a *closed-loop transfer function*. We will also have use for the transfer function, $E(s)$, relating the phase error, $\psi(t) = \phi(t) - \theta(t)$, to the input phase. This transfer function is

$$E(s) = \frac{\Psi(s)}{\Phi(s)} = \frac{\Phi(s) - \Theta(s)}{\Phi(s)} = 1 - \frac{\Theta(s)}{\Phi(s)} = 1 - H(s) \qquad (2.15)$$

Substitution of (2.14) into (2.15) gives

$$E(s) = \frac{\Psi(s)}{\Phi(s)} = 1 - H(s) = \frac{s}{s + GF(s)} \qquad (2.16)$$

2.2.1   PLL ORDER AND LOOP FILTERS

The acquisition and tracking behavior of a PLL depends in large measure upon the order of the loop, and for this reason, we consider a number of choices for the loop filter transfer function. The order of a PLL implementation is equal to the number of finite poles in the transfer function $H(s)$ given in (2.14). We will see that the order of a given PLL implementation exceeds the number of finite poles of $F(s)$ by one, with the additional pole resulting from the integration in the VCO model. Four loop filters are considered.

1. First-order PLL: The loop filter $F(s)$ in this case is a constant, say $F(s) = 1$, which results in a closed-loop transfer function of

$$H(s) = \frac{G}{s + G} \qquad (2.17)$$

Since there is only one pole in the transfer function, the order of PLL is one. Note that this single pole comes from the integration that models the VCO.

2. Second-order PLL with a perfect integrator: The transfer function of the loop filter is given by

$$F(s) = 1 + \frac{a}{s} \qquad (2.18)$$

which results in

$$H(s) = \frac{G(s+a)}{s^2 + Gs + Ga} \qquad (2.19)$$

3. Second-order PLL with an imperfect integrator: The transfer function of the loop filter is given by

$$F(s) = \frac{s+a}{s+\lambda a} \qquad (2.20)$$

which results in

$$H(s) = \frac{G(s+a)}{s^2 + (G+\lambda a)s + Ga} \qquad (2.21)$$

4. Third-order PLL with a perfect integrator: The transfer function of the loop filter is given by

$$F(s) = 1 + \frac{a}{s} + \frac{b}{s^2} \qquad (2.22)$$

which results in

$$H(s) = \frac{G(s^2 + as + b)}{s^3 + Gs^2 + Gas + Gb} \qquad (2.23)$$

2.2.2   STEADY-STATE PHASE ERRORS

Insight into the impact of using different loop filters can be obtained by observing steady-state values of the phase error for various input phase deviations. The final-value theorem from Laplace transform is very useful for this purpose[3]. From the final-value theorem, we know that a given time function, $a(t)$, has a steady-value (if the steady-state value exists) given by

$$\lim_{t \to \infty} a(t) = \lim_{s \to 0} s A(s) \qquad (2.24)$$

where $A(s)$ is the Laplace transform of $a(t)$.

The steady-state phase error is therefore given by

$$\Psi_{ss} = \lim_{s \to 0} s \Psi(s) = \lim_{s \to 0} s \Phi(s)[1 - H(s)] \qquad (2.25)$$

[3]See any standard textbook on linear control theory or any standard textbook on signal and system theory, for example (17).

Substitution of (2.16) then yields

$$\Psi_{ss} = \lim_{s \to 0} \Phi(s) \frac{s^2}{s + GF(s)} \qquad (2.26)$$

By substituting the appropriate function for $\Phi(s)$, the steady-state phase error is easily determined. In determining the steady-state phase errors, we assume an input phase deviation of the form

$$\phi(t) = \pi R t^2 + 2\pi \Delta_f t + \theta_0, \quad t > 0 \qquad (2.27)$$

The corresponding frequency deviation, in Hertz, is found from taking the derivative of (2.27) and is given by

$$\frac{1}{2\pi} \frac{d\phi}{dt} = Rt + \Delta_f \qquad (2.28)$$

For $R \neq 0$, the input phase deviation contains a quadratic phase term corresponding to a ramp in frequency. The quantity R is simply the slope of this ramp measured in Hertz per second. A nonzero Δ_f yields a frequency step of Δ_f Hertz, corresponding to a ramp in phase. The Laplace transform of (2.27) is

$$\Phi(s) = \frac{2\pi R}{s^3} + \frac{2\pi \Delta_f}{s^2} + \frac{\theta_0}{s} \qquad (2.29)$$

By substituting (2.29) into (2.26), the steady-state phase error becomes

$$\Psi_{ss} = \lim_{s \to 0} \left[\frac{2\pi R}{s^3} + \frac{2\pi \Delta_f}{s^2} + \frac{\theta_0}{s} \right] \frac{s^2}{s + GF(s)} \qquad (2.30)$$

for various choices of the loop filter transfer function $F(s)$ and for various conditions on θ_0, Δ_f and R.

The loop filter transfer functions of interest with the corresponding steady-state errors are shown in Table 2.1. Also given are the VCO phase to input phase transfer functions and the phase error to input phase transfer functions. Three cases of input phase defined by (2.29) are considered.

(1) $R = 0$, $\Delta_f = 0$, $\theta_0 \neq 0$: in this case, (2.30) becomes

$$\Psi_{ss} = \lim_{s \to 0} \frac{s\theta_0}{s + GF(s)} = 0 \qquad (2.31)$$

We can see that the steady-state phase error is zero for any type of loop filter transfer function, $F(s)$.

(2) $R = 0$, $\Delta_f \neq 0$, $\theta_0 \neq 0$: in this case, (2.30) becomes

$$\Psi_{ss} = \lim_{s \to 0} \left[\frac{2\pi \Delta_f}{s^2} + \frac{\theta_0}{s} \right] \frac{s^2}{s + GF(s)} = \frac{2\pi \Delta_f}{GF(0)} \qquad (2.32)$$

Table 2.1: Phase-Locked Loop Transfer Functions and Steady-state Errors Based on the Linear Model

Loop Type	Loop Filter Transfer Function $F(s)$	VCO Phase to Input Phase Transfer Function $\frac{\Theta(s)}{\Phi(s)} = H(s)$	Phase Error to Input Phase Transfer Function $\frac{\Psi(s)}{\Phi(s)} = 1 - H(s)$	Steady-State Error		
				$R = 0$ $\Delta_f = 0$ $\theta_0 \neq 0$	$R = 0$ $\Delta_f \neq 0$ $\theta_0 \neq 0$	$R \neq 0$ $\Delta_f \neq 0$ $\theta_0 \neq 0$
First order	1	$\dfrac{G}{s+G}$	$\dfrac{s}{s+G}$	0	$\dfrac{2\pi\Delta_f}{G}$	∞
Second order (Perfect)	$1 + \dfrac{a}{s}$	$\dfrac{G(s+a)}{s^2 + Gs + Ga}$	$\dfrac{s^2}{s^2 + Gs + Ga}$	0	0	$\dfrac{2\pi R}{Ga}$
Second order (Imperfect)	$\dfrac{s+a}{s+\lambda a}$	$\dfrac{G(s+a)}{s^2 + (G+\lambda a)s + Ga}$	$\dfrac{s^2 + \lambda a s}{s^2 + (G+\lambda a)s + Ga}$	0	$\lambda\dfrac{2\pi\Delta_f}{G}$	∞
Third order (Perfect)	$1 + \dfrac{a}{s} + \dfrac{b}{s^2}$	$\dfrac{G(s^2 + as + b)}{s^3 + Gs^2 + Gas + Gb}$	$\dfrac{s^3}{s^3 + Gs^2 + Gas + Gb}$	0	0	0

(2.a) For the first-order loop, $F(0) = 1$, the steady-state phase error is

$$\Psi_{ss} = \frac{2\pi \Delta_f}{G} \tag{2.33}$$

(2.b) For the perfect second-order loop, $F(s) = (s + a)/s$,

$$\Psi_{ss} = \lim_{s \to 0} \left[\frac{2\pi \Delta_f}{s^2} + \frac{\theta_0}{s} \right] \frac{s^3}{s^2 + G(s + a)} = 0 \tag{2.34}$$

(2.c) For the imperfect second-order loop, $F(0) = 1/\lambda$,

$$\Psi_{ss} = \lambda \frac{2\pi \Delta_f}{G} \tag{2.35}$$

(2.d) For the perfect third-order loop, $F(s) = (s^2 + as + b)/s^2$,

$$\Psi_{ss} = \lim_{s \to 0} \left[\frac{2\pi \Delta_f}{s^2} + \frac{\theta_0}{s} \right] \frac{s^4}{s^3 + G(s^2 + as + b)} = 0 \tag{2.36}$$

(3) $R \neq 0$, $\Delta_f \neq 0$, $\theta_0 \neq 0$: using a similar approach, the results can be obtained and are given in Table 2.1.

We can see, from Table 2.1, that to track an input frequency step ($\Delta_f \neq 0$) with zero steady-state error requires a perfect second-order loop. Tracking an input frequency ramp ($R \neq 0$) requires a third-order loop. It should be mentioned again that the steady-state phase errors given in Table 2.1 are based on a linear analysis, which requires that the phase error always be small. If this is not the case, the steady-state phase error can be substantially different, as we will see in later sections.

2.3   ACQUISITION AND PHASE PLANE ANALYSIS

2.3.1   FIRST-ORDER PLL

We now consider the first-order PLL for which

$$F(s) = 1 \tag{2.37}$$

or, equivalently,

$$f(t) = \delta(t) \tag{2.38}$$

Substitution of the above into (2.8) yields

$$\theta(t) = G \int_{-\infty}^{t} \int_{-\infty}^{\infty} \sin[\phi(\lambda) - \theta(\lambda)]\delta(\tau - \lambda)d\lambda d\tau \tag{2.39}$$

Performing the integration on λ using the sifting property of the delta function gives

$$\theta(t) = G \int_{-\infty}^{t} \sin[\phi(\tau) - \theta(\tau)] d\tau \tag{2.40}$$

and differentiating with respect to t gives

$$\frac{d\theta}{dt} = G \sin[\phi(t) - \theta(t)] \tag{2.41}$$

Since the phase error is $\psi(t) = \phi(t) - \theta(t)$, the frequency error can be written as

$$\frac{d\psi}{dt} = \frac{d\phi}{dt} - \frac{d\theta}{dt} = \frac{d\phi}{dt} - G \sin[\psi(t)] \tag{2.42}$$

We can now determine the phase error for a given input phase deviation. In order to study the response of a first-order PLL to a step in frequency of Δ_f Hertz at time t_o, we let

$$\frac{d\phi}{dt} = 2\pi \Delta_f u(t - t_o) \tag{2.43}$$

so that (2.42) becomes

$$\frac{d\psi}{dt} = 2\pi \Delta_f - G \sin[\psi(t)], \quad t > t_o \tag{2.44}$$

This yields a relationship between the frequency error and the phase error for $t > t_o$.

Equation (2.44), illustrated in Figure 2.7, defines the *phase plane* and describes the dynamic behavior of the first-order PLL with a frequency step input. The phase plane describes the manner in which the loop achieves lock. The phase plane has a number of important properties and understanding a few of them provides insight into how the loop achieves lock and the conditions under which phase lock will be achieved.

The relationship between phase error and frequency error must satisfy (2.44) at each point in time. These time dependent points are known as *operating points*. In the upper half phase plane ($d\psi/dt > 0$), the operating point moves from left to right, and in the lower half phase ($d\psi/dt < 0$), the operating point moves from right to left. This is easily seen. First, we let

$$\frac{d\psi}{dt} \approx \frac{\Delta\psi}{\Delta t} \tag{2.45}$$

where $\Delta\psi$ and Δt denote small increments in phase error and time, respectively. Clearly $\Delta t > 0$ for all t since time always increases. Thus $\Delta\psi > 0$ in the upper half phase plane and $\Delta\psi < 0$ in the lower half phase plane. The phase error $\psi(t)$ therefore increases in the upper half phase plane and decreases in the lower half phase plane. Stated another way, the operating point moves from

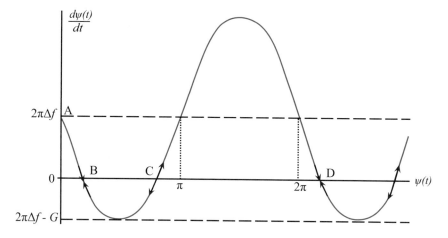

Figure 2.7: Phase Plane Plot for First-Order Loop as Defined by Eqn. (2.44)

left to right in the upper half phase plane and from right to left in the lower half phase plane.

The only way that the operating point can be stationary is for the operating point to lie on the boundary between the upper half phase plane and the lower half phase plane. This, of course, denotes that the phase error is constant or, equivalently, that the frequency error ($d\psi/dt$) is zero. The loop is locked only if the frequency error is zero. It can be seen from (2.44) that there are two points in each interval of 2π for which $d\psi/dt = 0$. If $d\psi/dt = 0$, $\sin[\psi(t)] = 2\pi\Delta_f/G$. Since $\sin[\psi(t)]$ cannot exceed 1, $2\pi\Delta_f \leq G$. In other words, the loop can lock only if $2\pi\Delta_f \leq G$.

An operating point is stable if, after a small perturbation, the operating point returns to original location. If a small perturbation results in the operating point moving to a new position, the original operating point is called unstable. Returning to Figure 2.7, points **B** and **C** are two points in the first period that $d\psi/dt = 0$. We can see that the point **B** is a stable operating point since, after a small perturbation, the operating point will return to point **B**. On the other hand, point **C** is an unstable point since a small perturbation will result in a movement of the operating point to **B** or **D**, depending upon the direction of the perturbation.

It can be seen from (2.44) that if $2\pi\Delta_f < G$ the steady-state operating point is the stable point **B** given that the initial point due to the frequency step is **A**. At this point, $d\psi/dt = 0$ and the steady-state phase error is the solution of

$$G\sin\psi_{ss} = 2\pi\Delta_f \tag{2.46}$$

Thus ψ_{ss} is given by

$$\psi_{ss} = \sin^{-1}\left[\frac{2\pi\Delta_f}{G}\right] \tag{2.47}$$

The value of ψ_{ss} for a first-order PLL with a step frequency input predicted by the linear model given by (2.30) is $2\pi\Delta_f/G$, which is smaller than the value given by (2.47). Note that the linear approximation becomes more accurate as Δ_f becomes smaller with respect to G since $\sin(\psi) \approx \psi$ for sufficiently small ψ. The following is an example in which the linear approximation does not accurately yield the steady-state error. Assume that $2\pi\Delta_f/G = 0.9$; the steady-state error is then given by

$$\psi_{ss} = \sin^{-1}\left[\frac{2\pi\Delta_f}{G}\right] = \sin^{-1}(0.9) = 1.19 \tag{2.48}$$

We can see that the linear approximation ($\psi_{ss} = 0.9$) underestimates the steady-state phase error by almost 25 percent.

A few final observations of the phase plane for the first-order PLL with a step frequency input are in order: If $2\pi\Delta_f < G$, there is a stable operating point in every cycle, which means that the phase cannot change by more than one cycle before it is locked. If $2\pi\Delta_f > G$, there is no solution to (2.44) for zero frequency error, and the operating point will move to the right for all time for $\Delta_f > 0$ and will move to the left for all time for $\Delta_f < 0$. The loop gain G therefore becomes the lock range for the first-order loop. The *lock range* is a frequency range over which the PLL can achieve a phase lock without slipping a cycle.

NUMERICAL EXAMPLE 2.1: Suppose that $G = 50$ and that we develop the phase plane for $2\pi\Delta_f = 40$ and 80 radians/second. The resulting phase planes[4] are shown in Figure 2.8.

The phase plane shows that for $2\pi\Delta_f < G$ the frequency error decreases to zero monotonically. There is no overshoot since the system is first order. On the other hand, for $2\pi\Delta_f > G$, there is no solution to (2.44). Therefore, PLL never locks.

The resulting input frequency deviation and VCO frequency deviation are shown in Figure 2.9(a) for $G = 50$ and $2\pi\Delta_f = 40$. Figure 2.9(b) illustrates the same characteristics for $2\pi\Delta_f > G$ ($2\pi\Delta_f = 80$ and $G = 50$), which shows that the VCO frequency deviation never reaches the value of input frequency deviation.

---

[4]All phase planes and other performance characteristics illustrated in the remainder of this chapter were generated by the simulation programs presented in the following chapters and are, of course, subject to the approximations made in the development of the simulation programs. See Chapter 4 for details on these approximations. All simulation programs can be found at http://www.morganclaypool.com/page/pll.

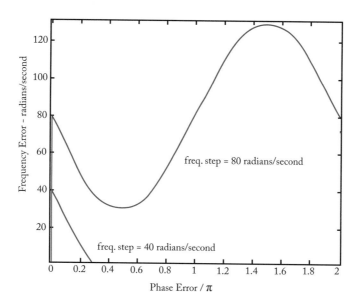

Figure 2.8: Phase Plane Plot for First-Order PLL

2.3.2   THE PERFECT SECOND-ORDER PHASE LOCK LOOP

The perfect second-order PLL is defined by the loop transfer function

$$F(s) = 1 + \frac{a}{s} = \frac{s+a}{s} \tag{2.49}$$

We call this a perfect loop because the pole is exactly at the origin, which denotes perfect integration. The perfect second-order PLL is defined by the loop transfer function

$$H(s) = \frac{G(\frac{s+a}{s})}{s + G(\frac{s+a}{s})} = \frac{G(s+a)}{s^2 + Gs + Ga} \tag{2.50}$$

Equating the denominator of the loop transfer function to the form of the denominator of the standard second-order linear system transfer function yields

$$s^2 + Gs + Ga = s^2 + 2\zeta(2\pi f_n)s + (2\pi f_n)^2 \tag{2.51}$$

where ζ is referred as the *damping factor* of the loop and f_n is the *natural frequency* of the loop expressed in Hertz.

In the simulation applications to follow, we will wish to define the loop damping factor and the loop natural frequency and then solve for the loop parameters that yield the required damping

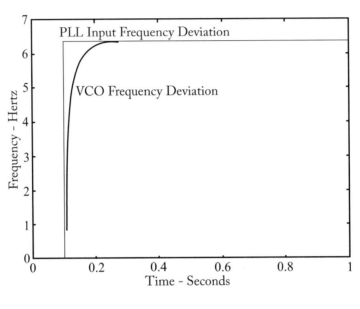

(a) $G = 50$ and $2\pi \Delta_f = 40$

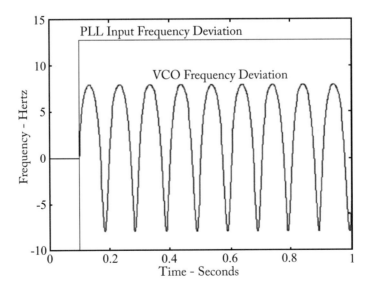

(b) $G = 50$ and $2\pi \Delta_f = 80$

Figure 2.9: Input and VCO Frequency Deviation

factor and natural frequency. Thus, given the damping factor and natural frequency, we set the loop gain

$$G = 4\pi \zeta f_n \qquad (2.52)$$

and the filter parameter

$$a = \frac{(2\pi f_n)^2}{G} = \frac{\pi f_n}{\zeta} \qquad (2.53)$$

It is, therefore, a very simple task to compute the loop gain and the filter parameter, which are required in the simulation, from the desired damping constant and natural frequency.

An advantage of using a perfect second-order PLL over a first-order PLL is that the lock range is infinite. In other words, the perfect second-order PLL will achieve phase lock with a frequency step on the input without regard to the size of the frequency step. It should be remembered, however, that the acquisition time may be excessive, and, therefore, some types of acquisition aid may be necessary[5]. It should also be remembered that a "perfect" second-order PLL cannot be exactly realized in practice since a perfect analog integrator is required. It is possible, however, to come very close for many applications.

NUMERICAL EXAMPLE 2.2: The phase plane plot for the perfect second-order PLL is illustrated in Figure 2.10. The particular case illustrated is Figure 2.10 was obtained using the simulation models developed in Chapter 4 with the following parameters:

input frequency step = 40 Hertz

loop natural frequency = 10 Hertz

loop damping factor = 0.707

The initial frequency error of 40 Hertz is clearly seen. It can also be seen that, after slipping three cycles, the loop achieves phase lock with zero steady-state phase error.

The input frequency deviation and the VCO frequency deviation corresponding to the phase plane shown in Figure 2.10 are illustrated in Figure 2.11. The cycle slipping behavior can now be seen in the time domain and it is clear that three cycles are slipped before the VCO frequency deviation reaches the input frequency deviation.

2.3.3   THE IMPERFECT SECOND-ORDER PHASE LOCK LOOP

The imperfect second-order PLL is defined by the loop filter transfer function

$$F(s) = \frac{s + a}{s + \lambda a} \qquad (2.54)$$

[5]See for example (1), (3), (6) for a discussion of acquisition aid.

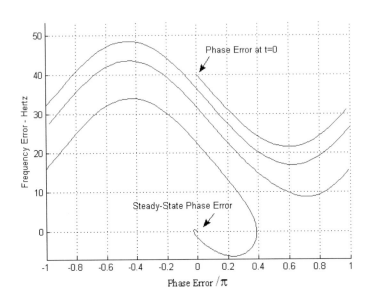

Figure 2.10: Phase Plane Plot for Perfect Second-Order PLL

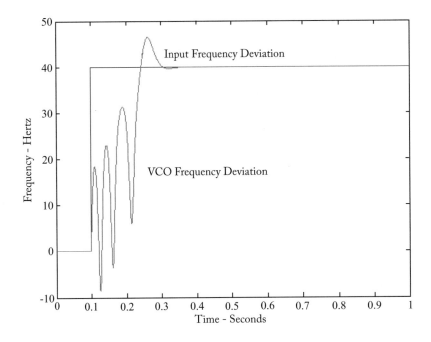

Figure 2.11: Input and VCO Frequency Deviation for Perfect Second-Order PLL

where λ denotes the offset of the pole from the origin relative to the zero location. Note that for $\lambda = 0$, the loop becomes "perfect" while for $\lambda = 1$, the loop becomes the first-order loop because of the pole-zero cancellation. In applications of the imperfect second-order loop, we will usually have interest in small values of λ.

The loop linear-model transfer function (2.21) is

$$H(s) = \frac{G(s + a)}{s^2 + (G + \lambda a)s + Ga} \tag{2.55}$$

It is often convenient to express the loop damping factor ζ^* and natural frequency f_n^* for the imperfect second-order loop in terms of ζ and f_n of the perfect second-order loop from which it is derived. In order to do this, we express the denominator of the preceding expression in the form

$$s^2 + (G + \lambda a)s + Ga = s^2 + 2\zeta^*(2\pi f_n^*)s + (2\pi f_n^*)^2 \tag{2.56}$$

The equivalent expression for the perfect second-order loop was expressed by (2.51) and is repeated here for reference:

$$s^2 + Gs + Ga = s^2 + 2\zeta(2\pi f_n)s + (2\pi f_n)^2 \tag{2.57}$$

From (2.56), we see that λ only appears in the middle term (the term which multiples s). The constant term (the term independent of s) establishes the natural frequency of the loop. It therefore follows that a non-zero value of λ has no effect on the loop natural frequency. Comparing (2.56) and (2.57), we see that

$$f_n^* = f_n \tag{2.58}$$

and a non-zero value of λ does affect the damping factor. From (2.56), we see that

$$4\pi f_n \zeta^* = G + \lambda a \tag{2.59}$$

Using the values of G and a that define the perfect second-order loop, (2.52) and (2.53), we get

$$4\pi f_n \zeta^* = 4\pi \zeta f_n + \lambda \frac{\pi f_n}{\zeta} \tag{2.60}$$

from which

$$\zeta^* = \zeta + \frac{\lambda}{4\zeta} \tag{2.61}$$

Thus, non-zero values of λ increase the damping factor. However, small values of λ have little effect on the damping factor unless the damping factor is small.

There are three main effects caused by non-zero values of λ. The most important of these effects is that for certain values of the frequency step Δ_f and pole offset factor λ the loop may not achieve phase lock but will exhibit limit cycle behavior. This effect is thoroughly analyzed by Viterbi (7). For situations in which the PLL will achieve phase lock, non-zero values of λ will tend to increase the lock time compared to the $\lambda = 0$ case. A non-zero steady-state error to an input frequency step also occurs, which increases with increasing λ as shown in Table 2.1.

NUMERICAL EXAMPLE 2.3: In order to show the effect of non-zero λ, we consider two cases: 1) one in which the time required to achieve phase lock is increased and 2) one in which limit cycle occurs.

For the first case, we consider the parameters used to develop Figure 2.10 but add a pole offset $\lambda = 0.2$. Thus, the simulation parameters are

input frequency step = 40 Hertz
loop natural frequency = 10 Hertz
loop damping factor = 0.707
pole offset $\lambda = 0.2$

The phase plane for these parameters is shown in Figure 2.12. We can see the initial frequency step of 40 Hertz and the non-zero steady state phase error. It can also be seen that the loop achieves phase lock after slipping fourteen cycles instead of three cycles when $\lambda = 0$ as shown in Figure 2.10.

We next consider the case in which the size of the input frequency step in increased from 40 Hertz to 50 Hertz. In other words, we consider the parameters

input frequency step = 50 Hertz
loop natural frequency = 10 Hertz
loop damping factor = 0.707
pole offset $\lambda = 0.2$

The phase plane for this case is illustrated in Figure 2.13. The limit cycle behavior is clearly seen, and the phase lock is never achieved.

2.3.4   THE PERFECT THIRD-ORDER PHASE LOCK LOOP

The perfect third-order PLL is defined by the loop filter transfer function

$$F(s) = 1 + \frac{a}{s} + \frac{b}{s^2} = \frac{s^2 + as + b}{s^2} \tag{2.62}$$

Once again, this is referred to as a perfect loop because the poles of the loop filter transfer function are exactly at the origin. The transfer function of a perfect third-order PLL (2.23) is

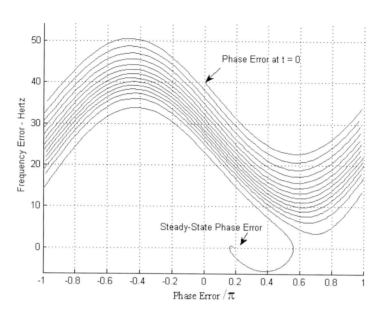

Figure 2.12: Phase Plane Plot for Imperfect Second-Order PLL

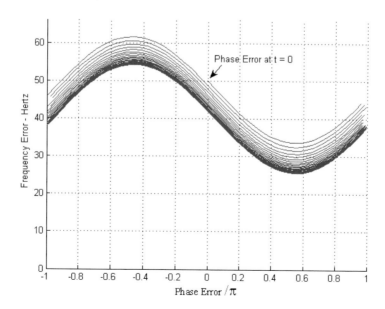

Figure 2.13: Phase Plane Plot for Imperfect Second-Order PLL showing Limit Cycle Behavior

$$H(s) = \frac{G(s^2 + as + b)}{s^3 + Gs^2 + Gas + Gb} \tag{2.63}$$

The third-order PLL is quite different from the first-order and second-order PLLs since a third-order PLL can be unstable for small values of loop gain, and, therefore, the third-order loop must be designed with great care. The advantage of using a third-order PLL lies in the fact that a third-order PLL can track a wider variety of signals compared with lower-order tracking loops. In a mobile communication system, when there is a relative motion between transmitting and receiving antennas, signals are subject to the well-known Doppler effect[6], which causes a frequency shift or a Doppler shift in the signal frequency. The Doppler shift depends upon the speed and direction of movement between the two antennas. The direction of movement can cause an increase or decrease in the signal frequency. As can be seen in Table 2.1, only the third-order PLL can suppress the steady-state phase error for the frequency ramp input ($R \neq 0$). Hence, the third-order PLL can be used in tracking the frequency ramp transition caused by the Doppler shift in a mobile communication system.

NUMERICAL EXAMPLE 2.4: With the PLL initially in lock, we apply an input frequency ramp. In other words, the input phase deviation is given by (2.29) with $\Delta_f = 0$ and $\theta_0 = 0$. The resulting phase plane is shown in Figure 2.14, which was created using the simulation models developed in Chapter 4 with the parameters:

$$\text{input frequency ramp} = 2500/\pi \text{ Hz/second}$$
$$\text{loop gain, } G = 100$$
$$\text{first filter parameter, } a = 50$$
$$\text{second filter parameter, } b = 2500$$

The resulting dynamics can clearly be seen. Both the initial frequency error and the final frequency error, as well as the initial and final phase errors, are zero. The frequency error initially increases rapidly, and then the loop achieves phase lock. Phase lock is achieved with zero steady-state error because of the assumed perfect integration.

2.3.5   TRANSPORT DELAY IN PHASE-LOCK LOOPS

Transport delay is a measure of the time required for the phase error signal to propagate around the loop to the VCO not counting the group delay of the loop filter, which is a part of the PLL design. Transport delay, which may be thought of as undesired delay imposed by a specific implementation, results from the fact that the loop components have a non-zero physical size and are separated by non-zero distances in the physical implementation of the device. If the carrier frequency is sufficiently high, this propagation time can be equivalent to a significant phase shift of the VCO

---

[6]Further discussion on the Doppler effect can be found in several wireless communication textbooks such as (18).

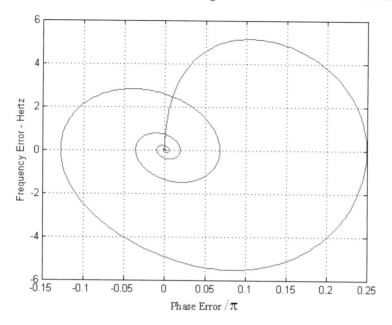

Figure 2.14: Phase Plane for a Third-Order PLL with a Ramp Frequency Input

output signal relative to the input signal.

Transport delay changes the linear model open loop transfer function from the design value given by $GF(s)/s$ to $[GF(s)/s]e^{-sT_d}$ where T_d is the transport delay. This illustrates the fact that the transport delay induces an additional phase shift of $2\pi f T_d$ radians on the open loop transfer function. This effect is, of course, obvious since a time delay in the time domain is equivalent to a phase shift in the frequency domain. This extra phase shift arising from transport delay reduces the phase margin of the PLL and drives the PLL towards instability. This effect will be seen in the simulation to follow and in Chapter 4. It is often said that the effect of transport delay is to induce extra poles in the closed loop transfer function of the PLL (19). In the presence of significant transport delay, this additional phase shift cannot be neglected, and the linear model closed loop transfer function is changed from the design value defined by (2.14) to

$$H_{td}(s) = \frac{GF(s)e^{-sT_d}}{s + GF(s)e^{-sT_d}} \qquad (2.64)$$

which can be written

$$H_{td}(s) = \frac{GF(s)}{se^{sT_d} + GF(s)} \qquad (2.65)$$

Expanding e^{sT_d} in a power series clearly illustrates the source of the poles induced by transport delay.

NUMERICAL EXAMPLE 2.5: In order to briefly show the effect of transport delay, the result of one simulation using the Perfect Second-Order PLL is presented in Figure 2.15. The following parameters were used:

$$\text{input frequency step} = 40 \text{ Hertz}$$
$$\text{loop natural frequency} = 10 \text{ Hertz}$$
$$\text{loop damping factor} = 0.707$$
$$\text{pole offset } \lambda = 0 \text{ (perfect integration)}$$
$$\text{transport delay} = 9 \text{ sample periods}[7]$$

Since the default sampling frequency of 2000 Hertz was used, the transport delay is 0.0045 seconds. In order to see the effect of this delay, one should compare Figure 2.15 with Figure 2.10, which shows the simulation results without the effect of transport delay. From Figure 2.10, we see that without transport delay, the PLL slips three cycles, while Figure 2.15 shows that with a transport delay of 9 sample periods, the PLL slips 9 cycles prior to achieving phase lock. The resulting time required to achieve phase lock is therefore increased substantially. Phase lock is eventually achieved, however, with zero steady-state error. It turns out that if the transport delay is increased from 9 sample periods to 11 sample periods using the same loop parameters, the PLL will never achieve phase lock. In studying Figure 2.15, one should note the initial horizontal line at a value of 40 Hertz frequency error. The fact that this line is horizontal shows that it takes a non-negligible time for the loop to start responding to the initial frequency step. This results from the transport delay and phase error accumulates over this time.

In Chapter 4, we will see that transport delay is modeled by placing a tapped delay line between the loop filter and the VCO. The simulations given in Chapter 4 can be used for gaining additional insight into the relationship between transport delay and loop dynamics.

2.4 PROBLEMS

2.1 Run a simulation for a first-order PLL to confirm the results given in Figures 2.8 and 2.9.

2.2 In Problem 2.1, what is the steady-state phase error mod(2π) for $G = 50$? Using the linear model (Table 2.1), find the steady-state phase error and compare it with the phase error from the simulation. Repeat for $G = 100$. Explain the effect of G.

---

[7]Even though the transport delay is set equal to 9 sample periods, the true transport delay for this simulation is 10 sample periods. The excess delay of one sample period is caused by the manner in which the simulation is developed using the fixed-step trapezoid integration algorithm. This effect is discussed in Section 4.5.

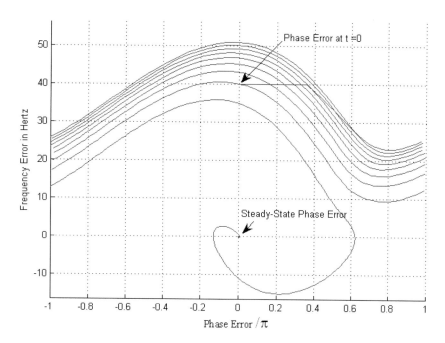

Figure 2.15: Second-Order PLL Dynamics with Transport Delay

2.3 Using the PLL linear phase model for the first-order PLL (2.17) and the Laplace transform, show that the transient responses of the VCO output phase and the phase error for a frequency step input $\Phi(s) = 2\pi \Delta_f / s^2$ are

$$\theta(t) = 2\pi \Delta_f [t - \frac{1}{G}(1 - e^{-Gt})]$$
$$\psi(t) = \frac{2\pi \Delta_f}{G}[1 - e^{-Gt}]$$

Also show that the value of $\psi(t)$ as $t \to \infty$ is equal to the value in (2.33).

2.4 From Problem 2.3, find the VCO frequency deviation $d\theta/dt$. Using $2\pi \Delta_f = 40$ rad/sec and $G = 50$, plot the VCO frequency deviation for t=[0:0.01:1] seconds and compare it with Figure 2.9(a). Note that your plot should start at t = 0.1 seconds.

2.5 Repeat Problem 2.4 for $2\pi \Delta_f = 80$ rad/sec and $G = 50$. Compare it with Figure 2.9(b) and explain what causes the difference in those two plots.

2.6 Using a simulation for the perfect second-order PLL with the same parameters given in Numerical Example 2.2 except the loop damping factor, find the numbers of cycle slips and the steady-state phase errors for the following values of the loop damping factor. Explain the effect of the loop damping factor. Note: use the extended phase plane plot to fill in the table below.

Loop damping factor	0.2	0.4	0.6	0.707	0.8	1.0
No. of cycle slips				3		
Steady-state phase errors				6π		

2.7 Given that the first-order PLL is used as a FM demodulator where the FM input signal is $m(t) = Au(t)$, the PLL input signal is therefore

$$x_{in}(t) = A_c\cos[2\pi f_c t + \phi(t)] = A_c\cos\left[2\pi f_c t + m_f A \int u(\lambda)d\lambda\right]$$

Using linear analysis and the Laplace transform, show that $e_{vco}(t)$ is a demodulated output and is given as

$$e_{vco}(t) = \frac{Am_f}{2\pi K_d}(1 - e^{-Gt})u(t)$$

Note that for $t >> 1/G$ and $m_f = 2\pi K_d$, $e_{vco}(t) = Au(t)$ which is the FM input signal.

CHAPTER 3

Structures Developed From The Basic PLL

In this chapter, we develop several of the structures that are useful in communications applications and based upon the basic PLL. While there are a number of candidate devices that we could explore in this chapter, we restrict our attention to the most common devices that form the basic building blocks for more complex systems. These include the Costas PLL, the QPSK tracking loop and the N-Phase tracking loop.

3.1 THE COSTAS PHASE-LOCKED LOOP

The Costas PLL is illustrated in Figure 3.1. It is useful in both analog and in digital communications systems. The Costas PLL can be used as a demodulator for either DSB/SC (double-sideband suppressed-carrier) signals in analog communications systems or BPSK (binary phase-shift keyed) signals in digital communications systems. The focus of the application developed here is on the development of a BPSK demodulator.

The input signal is assumed to have the form

$$x_{in}(t) = A_c m(t) \cos[2\pi f_c t + \phi(t)] \tag{3.1}$$

where $m(t)$ represents the information bearing, or message, signal and $\phi(t)$ represents the undesired phase deviation of the carrier.

For the case of interest, the BPSK signals, we assume that the message signal $m(t)$ is represented by the NRZ (non-return to zero) waveform having levels +1 or -1, depending upon the state of the information-bearing bit in the signaling interval of interest. The undesired phase deviation, $\phi(t)$, may be from phase jitter of the carrier at the transmitter arising from oscillator instabilities, from channel impairments, or may arise from a variety of other sources. The demodulator has the task of tracking this phase deviation so that the information-bearing signal can be recovered with the smallest possible error. This can be accomplished with a Costas PLL.

The signals present at various points in the loop are shown in Figure 3.1. In determining the various signals, we have assumed that the VCO output consists of two unity-amplitude sinusoids that are in phase quadrature. It is also assumed that the lowpass filters have a passband gain of 2

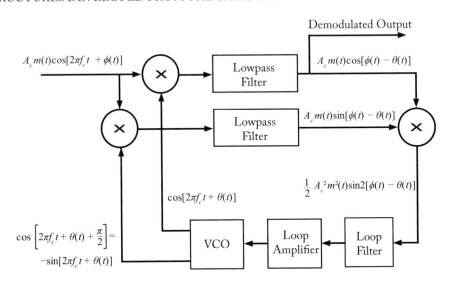

Figure 3.1: Costas Phase-Locked Loop

in order to remove the factor of 0.5, resulting from the trigonometric identity for $\cos(x)\cos(y)$ and $\cos(x)\sin(y)$. The output of the top lowpass filter is known as the direct channel output and the output of the bottom lowpass filter is known as the quadrature channel output. In other words, the direct channel output is given by

$$y_d(t) = A_c m(t)\cos[\phi(t) - \theta(t)] \tag{3.2}$$

and the quadrature channel output is given by

$$y_q(t) = A_c m(t)\sin[\phi(t) - \theta(t)] \tag{3.3}$$

Since the direct channel signal is proportional to the message signal, the demodulated output is usually considered to be the direct channel signal as shown in Figure 3.1. The loop control signal is the product of the direct channel and quadrature channel signals. Thus

$$e(t) = y_d(t)y_q(t) = \frac{1}{2}A_c^2 m^2(t)\sin\left[2[\phi(t) - \theta(t)]\right] \tag{3.4}$$

As stated earlier, we assume that the message signal $m(t)$ is represented by the NRZ (non-return to zero) waveform having levels +1 or -1, depending upon the state of the information-bearing bit in the signaling interval of interest. If $m(t)$ is +1 or -1, $m^2(t) = 1$ and the loop control signal becomes

$$e(t) = y_d(t)y_q(t) = \frac{1}{2}A_c^2 \sin\left[2[\phi(t) - \theta(t)]\right] \tag{3.5}$$

which, except for the fact that the phase error is multiplied by two in the argument of the sine function, is equivalent to the loop control signal in the basic PLL. It follows, therefore, that the operating characteristics of the Costas PLL are very similar to the operating characteristics of the basic PLL examined in the previous chapter.

Several points are worth noting. First, if the loop is operating in the region of phase lock so that the phase error is small, the direct channel signal is

$$y_d(t) = A_c m(t) \cos[\phi(t) - \theta(t)] \approx A_c m(t) \tag{3.6}$$

Since this is proportional to $m(t)$, it follows that the demodulated output is the direct channel signal as shown in Figure 3.1. For the case of small phase error, the quadrature signal can be approximated

$$y_q(t) = A_c m(t) \sin[\phi(t) - \theta(t)] \approx A_c m(t)[\phi(t) - \theta(t)] \tag{3.7}$$

The loop control signal for the small phase error case becomes

$$e(t) = y_d(t) y_q(t) = \frac{1}{2} A_c^2 \sin[2[\phi(t) - \theta(t)]] \approx A_c^2[\phi(t) - \theta(t)] \tag{3.8}$$

which shows that the loop control signal is proportional to the phase error. This was exactly the case with the basic PLL. As a matter of fact, the constant A_c^2 can be lumped into the gain of the loop amplifier. Thus, for operation in the region of phase lock, we may consider the loop control signal to be

$$e(t) \approx \phi(t) - \theta(t) \tag{3.9}$$

This is identical to the loop control signal in the linear model of the basic PLL. It is important to recognize that, for all practical purposes, it is the quadrature channel signal that maintains the loop in phase lock once phase lock is initially achieved.

The preceding paragraph illustrates that in the region of phase lock, the output of the direct channel simply multiplies the quadrature channel signal by a constant. If this constant is lumped into the loop gain, the direct channel is eliminated, and the Costas PLL reduces to the basic PLL discussed in the previous Chapter. It therefore follows that the loop filters used to establish the loop dynamics of the basic PLL are appropriate for application to the Costas PLL, and the filters given in Table 2.1 are therefore used for the Costas PLL simulations. It should also be noted that the linear model transfer functions given in Table 2.1 are also valid for the Costas PLL. Care should be used, however, to insure that the gain of the loop filter is correctly determined.

Another important observation follows from (3.5). Since the phase error is multiplied by 2 in the argument of the sine function, phase errors of the form

$$\psi(t) = \phi(t) - \theta(t) \pm \pi \tag{3.10}$$

cannot be distinguished from phase errors of the form

$$\psi(t) = \phi(t) - \theta(t) \tag{3.11}$$

It therefore follows that the Costas PLL has a π radian phase ambiguity. It is therefore possible for the demodulated output to be $-m(t)$ rather than $m(t)$ as desired. If this is a problem in a given application, the phase ambiguity must be removed external to the demodulation loop. Differential coding can be used for this purpose.

The linear model of the Costas PLL is easily developed from observation of the small phase error approximations expressed by (3.6), (3.7) and (3.8). Absorbing the constant A_c^2 into the gain of the loop amplifier and assuming that $m^2(t) = 1$ results in the linear model shown in Figure 3.2. Since the loop control signal is simply the sine of the phase error, the model shown in Figure 3.2 is equivalent to the basic PLL. It should be noted that in the model illustrated in Figure 3.2, the loop input and the VCO output are the complex envelopes of the signals present in the physical device. The complex envelope signals representations and phase detector models are discussed in further details in Appendix A. In addition, as previously noted, the loop filters used in the basic PLL and discussed in the previous chapter can therefore be used and equivalent behavior will result.

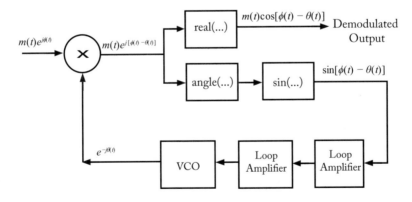

Figure 3.2: Linear Model of the Costas PLL for $m^2(t) = 1$

3.2   THE QPSK LOOP

The QPSK loop, illustrated in Figure 3.3, is a modification of the Costas PLL (9), (10). The development of the QPSK loop is well beyond the scope of this book[1]. We shall simply present the

---

[1]The QPSK loop is developed based on the maximum a posteriori probability (MAP) approach. The hard-limiter function is used in place of a tanh(.) function in an original design for an implementation purpose. It has been shown that the implementation with hard-limiters performs well in most conditions especially when SNR is high. Note that $\tanh(x)$ can be approximated by $sgn(x)$ or x when magnitude of x is large or small, respectively. See (10) for details.

loop structure and then, in the following chapter, develop a simulation model based on this loop structure.

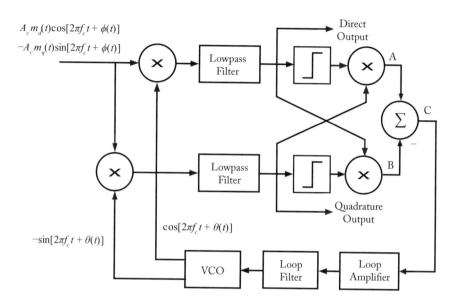

Figure 3.3: QPSK Loop

The input signal is assumed to have the form

$$x_{in}(t) = A_c m_d(t)\cos[2\pi f_c t + \phi(t)] - A_c m_q(t)\sin[2\pi f_c t + \phi(t)] \tag{3.12}$$

where $m_d(t)$ and $m_q(t)$ represent the information bearing, or message, signals in the direct channel and the quadrature channel, respectively, and $\phi(t)$ represents the undesired phase deviation of the carrier.

Similar to the BPSK demodulation, in Figure 3.3, we have assumed that the VCO output consists of two unity-amplitude sinusoids that are in phase quadrature. It is also assumed that the lowpass filters have a passband gain of 2. The direct channel output is given by

$$y_d(t) = A_c m_d(t)\cos[\phi(t) - \theta(t)] - A_c m_q(t)\sin[\phi(t) - \theta(t)] \tag{3.13}$$

and the quadrature channel output is given by

$$y_q(t) = A_c m_d(t)\sin[\phi(t) - \theta(t)] + A_c m_q(t)\cos[\phi(t) - \theta(t)] \tag{3.14}$$

Since $\psi(t) = \phi(t) - \theta(t)$, the signal at **A** in Figure 3.3 can be written as

$$\text{sgn}\{y_d(t)\}\, y_q(t) = \text{sgn}\{y_d(t)\}\left\{A_c m_d(t)\sin[\psi(t))] + A_c m_q(t)\cos[\psi(t)]\right\} \tag{3.15}$$

and the signal at **B** in Figure 3.3 can be written as

$$\text{sgn}\left\{y_q(t)\right\} y_d(t) = \text{sgn}\left\{y_q(t)\right\}\left\{A_c m_d(t)\cos[\psi(t)] - A_c m_q(t)sin[\psi(t)]\right\} \quad (3.16)$$

where sgn {...} is a signum function representing a hard limiter operation.

The loop control signal, signal at **C**, is given as (3.15)-(3.16):

$$e(t) = \text{sgn}\left\{y_d(t)\right\} y_q(t) - \text{sgn}\left\{y_q(t)\right\} y_d(t) \quad (3.17)$$

Given that the message signals $m_d(t)$ and $m_q(t)$ are represented by the NRZ waveform having levels +1 or −1, depending upon the state of the information-bearing bit, the following four cases of $\psi(t)$ given in Table 3.1 are considered for the loop control signal, $e(t)$.

Table 3.1: Loop Control Signal, $e(t)$					
$\psi(t)$	$m_d(t)$	$m_q(t)$	$\text{sgn}\left\{y_d(t)\right\}$	$\text{sgn}\left\{y_q(t)\right\}$	$e(t)$
$-\frac{\pi}{4} < \psi(t) \leq \frac{\pi}{4}$	+1	+1	+1	+1	$2A_c\sin[\psi(t)]$
	+1	−1	+1	−1	$2A_c\sin[\psi(t)]$
	−1	+1	−1	+1	$2A_c\sin[\psi(t)]$
	−1	−1	−1	−1	$2A_c\sin[\psi(t)]$
$\frac{\pi}{4} < \psi(t) \leq \frac{3\pi}{4}$	+1	+1	−1	+1	$-2A_c\cos[\psi(t)]$
	+1	−1	+1	+1	$-2A_c\cos[\psi(t)]$
	−1	+1	−1	−1	$-2A_c\cos[\psi(t)]$
	−1	−1	+1	−1	$-2A_c\cos[\psi(t)]$
$\frac{3\pi}{4} < \psi(t) \leq \frac{5\pi}{4}$	+1	+1	−1	−1	$-2A_c\sin[\psi(t)]$
	+1	−1	−1	+1	$-2A_c\sin[\psi(t)]$
	−1	+1	+1	−1	$-2A_c\sin[\psi(t)]$
	−1	−1	+1	+1	$-2A_c\sin[\psi(t)]$
$\frac{5\pi}{4} < \psi(t) \leq \frac{7\pi}{4}$	+1	+1	+1	−1	$2A_c\cos[\psi(t)]$
	+1	−1	−1	−1	$2A_c\cos[\psi(t)]$
	−1	+1	+1	+1	$2A_c\cos[\psi(t)]$
	−1	−1	−1	+1	$2A_c\cos[\psi(t)]$

Several points are worth noting. First, if the loop is operating in the region of phase lock so that the phase error is small and closes to zero, the direct channel signal is

$$y_d(t) = A_c m_d(t)\cos[\phi(t) - \theta(t)] - A_c m_q(t)\sin[\phi(t) - \theta(t)] \approx A_c m_d(t) \quad (3.18)$$

which is proportional to the demodulated output of the direct channel input. The quadrature signal can be approximated as

$$y_q(t) = A_c m_d(t)\sin[\phi(t) - \theta(t)] + A_c m_q(t)\cos[\phi(t) - \theta(t)] \approx A_c m_q(t) \quad (3.19)$$

which is proportional to the demodulated output of the quadrature channel input. The loop control signal for the small phase error case becomes

$$e(t) = 2A_c\sin[\psi(t)] \approx 2A_c[\phi(t) - \theta(t)] \tag{3.20}$$

which shows that the loop control signal is proportional to the phase error. This was exactly the case with the basic PLL. As a matter of fact, the constant $2A_c$ can be lumped into the gain of the loop amplifier. Thus, for operation in the region of phase lock, we may consider the loop control signal to be

$$e(t) \approx \phi(t) - \theta(t) \tag{3.21}$$

which is simply the phase error.

Second, we can see in Table 3.1 that the loop control signal $e(t)$ is a function of a phase error $\psi(t)$ and is illustrated in Figure 3.4. In addition, the loop control signal is a periodic function of ψ with a period of $\frac{\pi}{2}$. Using Fourier series, the loop control signal can also be written as[2]

$$e(\psi) = -\frac{64A_c}{\sqrt{2}\pi} \sum_{n=1}^{\infty} \frac{n\cos(n\pi)}{16n^2 - 1}\sin(4n\psi) \tag{3.22}$$

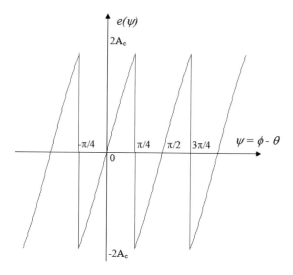

Figure 3.4: Loop Control Signal vs. ψ

In discussing the Costas PLL, we made note of the fact that the Costas PLL exhibits a π radian phase ambiguity and that the output may be negated because the loop lacks π radian out of

[2]Note that the dependence on t is implicit.

phase with the input signal. In other words, there are two stable operating points for each cycle of the input. The QPSK loop has a similar behavior. There are four stable operating points for each cycle of the input. Since the phase error, consider (3.22) when n = 1, is multiplied by 4 in the argument of the sine function, phase errors of the form $\psi(t) = \phi(t) - \theta(t) \pm \pi/4$ cannot be distinguished from phase errors of the form $\psi(t) = \phi(t) - \theta(t)$. It therefore follows that the QPSK loop has a $\pi/2$ radian phase ambiguity. The four different possibilities can be seen from Figure 3.5.

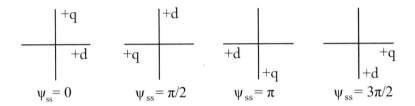

Figure 3.5: Phase Ambiguities for QPSK Loop

With ψ_{ss} representing the steady-state error, from (3.13) and (3.14), we have

$$y_d(t) = A_c m_d(t)\cos[\phi(t) - \theta(t)] - A_c m_q(t)\sin[\phi(t) - \theta(t)]$$

and

$$y_q(t) = A_c m_d(t)\sin[\phi(t) - \theta(t)] + A_c m_q(t)\cos[\phi(t) - \theta(t)],$$

one can visualize the coordinate system for $\psi_{ss} = 0$ to represent the basis set for the transmitted signal space. It also represents the basis set for the output for the case in which the QPSK loop achieves phase lock with a steady-state phase error of zero degrees (mod 2π). If the QPSK loop locks with $\psi_{ss} = \pi/2$, the direct channel output of the receiver $y_d(t)$ is the negative of the quadrature channel signal $-m_q(t)$, and the receiver quadrature output signal $y_q(t)$ is the direct channel signal $m_d(t)$. If $\psi_{ss} = \pi$, the direct and quadrature channel signals appear on the receiver outputs, but both are negated. This reminds us the Costas PLL with $\psi_{ss} = \pi$. For $\psi_{ss} = 3\pi/2$, the direct and quadrature channel signals are once again interchanged and the quadrature channel signal is negated. The four possibilities are summarized in Table 3.2 and illustrated in Figure 3.5.

A simulation of the QPSK demodulation loop is illustrated in Chapter 5. For the chosen simulation parameters, we will see that the result shows a steady-state phase error of 90 degrees and that the behavior predicted in Table 3.2 is illustrated.

3.3   THE *N*-PHASE TRACKING LOOP

The *N*-Phase tracking loop, sometimes referred to as the *N*-Phase Costas loop, is a generalization of the Costas PLL illustrated in Figure 3.1. With a little care, it is easily seen that the *N*-Phase

Table 3.2: Results of QPSK Loop Phase Ambiguities

Steady-State Phase Error Ψ_{ss}	Direct Receiver Output $y_d(t)$	Quadrature Receiver Output $y_q(t)$
0	$m_d(t)$	$m_q(t)$
$\pi/2$	$-m_q(t)$	$m_d(t)$
π	$-m_d(t)$	$-m_q(t)$
$3\pi/2$	$m_q(t)$	$-m_d(t)$

Costas loop is equivalent to the Costas PLL for $N = 2$. A nice development of the N-Phase tracking loop is presented in the book by Lindsey and Simon, in which the N-Phase tracking loop is shown to be equivalent to an Nth order power loop (8).

The N-Phase tracking loop is illustrated in Figure 3.6. Note that in this case, there are N VCO outputs given by

$$x_{i,vco}(t) = \cos\left[2\pi f_c t + \theta(t) - \frac{i\pi}{N}\right], \quad i = 0, 1, 2, ..., N-1 \tag{3.23}$$

where we have once again set the peak amplitude of all N VCO outputs equal to unity for mathematical convenience. In writing the equations for the signals present at various points in the loop, any amplitude variation on the loop input signal is neglected. Each VCO output multiplies the input signal, and the result of the multiplication is lowpass filtered. This operation yields N lowpass filter outputs of the form

$$y_i(t) = \cos\left[\phi(t) - \theta(t) + \frac{i\pi}{N}\right], \quad i = 0, 1, 2, ..., N-1 \tag{3.24}$$

These can be viewed as the N demodulated outputs. These N lowpass filter outputs are then multiplied together. The output of the multiplier is given by

$$y(t) = \prod_{i=0}^{N-1} \cos\left[\phi(t) - \theta(t) + \frac{i\pi}{N}\right] \tag{3.25}$$

For certain values of N, namely values of N which are a multiple of 4, the preceding expression can be written

$$y(t) = \frac{1}{2^{N-1}} \sin\left[N[\phi(t) - \theta(t)]\right] \tag{3.26}$$

For N-PSK systems, in which there are $b = \log_2 N$ bits of information transmitted per symbol, N is indeed a multiple of 4, and so (3.26) can be used for (3.25). It should be remembered, however, that (3.25) and (3.26) do not, in general, constitute a trigonometric identity.

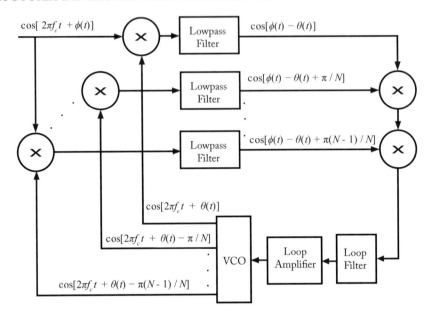

Figure 3.6: N-Phase Tracking Loop

The signal $y(t)$ is assumed to pass through an amplifier with gain

$$G_1 = \frac{2^{N-1}}{N} \tag{3.27}$$

in order to form the error control signal for the loop. The error control signal then becomes

$$e(t) = \frac{1}{N} \sin\left[N[\phi(t) - \theta(t)]\right] \tag{3.28}$$

If the argument of the sine function is sufficiently small, we then get $e(t) \approx \phi(t) - \theta(t)$ which is equivalent to the error control signal for a Costas PLL as expressed by (3.9) and to the error control signal for a basic PLL operating in the near lock condition. The gain expressed by (3.27) is not generally explicitly expressed, and therefore it is not shown in Figure 3.6. It is usually combined with the gain of the loop amplifier. It can be seen from (3.28) that the N-Phase tracking loop has N phase ambiguities per cycle. Similar behavior was noticed with the Costas PLL for which $N = 2$. These ambiguities must be resolved in most applications.

3.4   PROBLEMS

3.1 List three other applications of the PLL and briefly explain how PLL is used in each application.

3.2 Using MATLAB codes for the Costas phase-locked loop with the following parameters:

Input frequency step in Hertz > 40
Input loop natural frequency in Hertz > 20
Input samples per bit (should be a multiple of five) > 20

Plot a) the phase plane plot and b) the direct channel output and the message $m(t)$. What is the steady-state phase error mod(2π)? How long does it take for the Costas loop to regain phase lock. After reacquiring phase lock, state the relationship between the message $m(t)$ and the direct channel output.

3.3 Continued from Problem 3.2, change the input frequency step to 30 Hertz and keep the other parameters the same. Plot a) the phase plane plot and b) the direct channel output and the message $m(t)$. What is the steady-state phase error mod(2π)? How long does it take for the Costas loop to regain phase lock? After reacquiring phase lock, state the relationship between the message $m(t)$ and the direct channel output.

3.4 Write a MATLAB program to show that $e(t)$ in Table 3.1 and $e(t)$ in (3.22) yield the same graph as shown in Figure 3.4.

3.5 Write a MATLAB program to show that $y(t)$ in both (3.25) and (3.26) yield the same graph for $N = 4$. Repeat for $N = 8$. Also show that the graphs from both equations are different for value of N that is not an integer multiple of 4.

CHAPTER 4

Simulation Models

Simulation models for the structures presented in Chapter 2 are now presented. The trapezoidal-rule integration algorithm was used in the realizations of the VCO and loop filters. Trapezoidal integration and the realization of the various loop filters is discussed at length in Appendix B[1]. The simulation code can be found at `http://www.morganclaypool.com/page/pll` and the contents are listed in Appendix D. In the following sections, the signals present at each node in the system (**phin, phivco, s1, s2, ...**)[2] are identical to the corresponding variable names in the MATLAB simulation codes.

4.1 BASIC MODELS FOR PHASE-LOCKED LOOPS

Six different simulation models have been developed for the basic PLL structures. All simulation code is contained in a script file with the name **pllxyyy** where **x** denotes the order of the loop (1, 2, or 3) and **yyy** describes the loop phase detector or the presence of transport delay. The script files are defined as follows:

pll1sin.m	First-order PLL with sinusoidal phase detector
pll2sin.m	Second-order PLL with sinusoidal phase detector
pll2tri.m	Second-order PLL with triangular-wave phase detector
pll2saw.m	Second-order PLL with sawtooth-wave phase detector
pll2tdel.m	Second-order PLL with sinusoidal phase detector and transport delay
pll3sin.m	Third-order PLL with sinusoidal phase detector

There are obviously other examples that could be presented, but these are the most representative and correspond to the examples discussed in Chapter 2. In all of the models mentioned above, a number of integrations are necessary to build the simulation model. There are N (N =1, 2 or 3) integrations in each model with 1 integration used for realization of the VCO and $N - 1$ integrations used in the realization of the loop filter.

The simulation model for the first-order PLL is shown in Figure 4.1. This model corresponds to the MATLAB m-file **pll1sin.m**. The phase detector is sinusoidal although this is easily changed if desired. The VCO is represented by an integration process. As mentioned earlier, trapezoidal

---

[1]Also see Tranter, et.al [18] for further discussion of these techniques.
[2]Throughout this chapter, MATLAB variable names appear in boldface.

integration is used to model the integration.

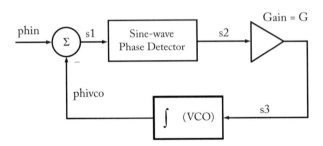

Figure 4.1: First-order PLL Simulation Model

The simulation model for the second-order PLL is shown in Figure 4.2 in which the form of the phase detector is not specified. There are three phase detectors available; sine-wave, triangular-wave and sawtooth-wave. The simulation code for these phase detectors are contained in the MAT-LAB m-files **pll2sin.m**, **pll2tri.m** and **pll2saw.m**, respectively. Note that the form of the phase detector depends on either the input waveforms applied to the multiplier of the phase detector used in Chapter 2 (Figure 2.3) or using other types of phase detector such as EXOR digital phase detectors which yield the triangular-wave characteristic and JK flip-flop digital phase detectors which yield the sawtooth-wave characteristic (1), (4). For example, if both inputs of the multiplier phase detector are both rectangular waveforms, the characteristic of the phase detector becomes triangular.

The m-files **pll2sin.m**, **pll2tri.m** and **pll2saw.m** are identical except for the phase detector models. Note that for each of the three models the input to the block defining the phase detector characteristic is the scaler **s1** and the output is the scaler **s2**. For the sinusoidal-wave phase detector, the nonlinear characteristic is implemented simply by the expression

$$s2 = \sin(s1) \tag{4.1}$$

as can be seen in the m-file **pll2sin.m**.

The triangular-wave phase detector is implemented using a Fourier series approximation to the triangular-wave phase detector characteristic. Using a six-term approximation leads to the expression

$$s2 = \frac{4}{\pi} \sum_{n=0}^{5} b_n \sin[(2n+1)s1] \tag{4.2}$$

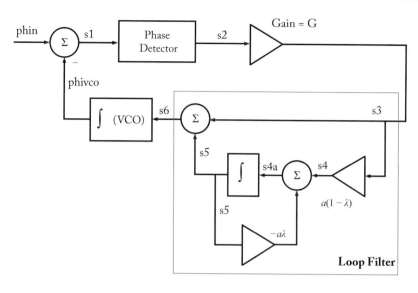

Figure 4.2: Second-order PLL Simulation Model

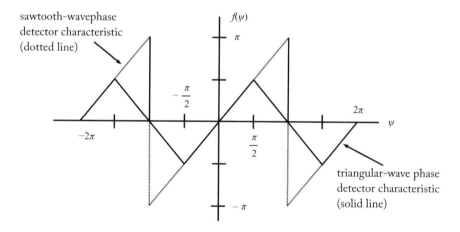

Figure 4.3: Triangular-wave and Sawtooth-wave Phase Detector Characteristics

in which b_n represents the Fourier coefficients given by

$$b_n = \frac{(-1)^n}{(2n+1)^2} \tag{4.3}$$

The decision to use a six-term approximation to the triangular-wave characteristic is arbitrary, and it is interesting to experiment with approximations of varying length. Equations (4.2) and (4.3) are realized using the MATLAB expression

$$s2 = (4/pi)^*sin([\ 1 \quad 3 \quad 5 \quad 7 \quad 9 \quad 11 \ ]^*s1)^*b' \tag{4.4}$$

in which **b** is the vector of Fourier coefficients

$$b = [\ 1 \quad -1/9 \quad 1/25 \quad -1/49 \quad 1/81 \quad -1/121 \ ] \tag{4.5}$$

as defined by (4.3). Note that since **b** is a vector of constants, it is defined prior to execution of the simulation loop as can be seen in the m-file **pll2tri.m**. Fourier series modeling techniques are important since they can be used to represent any memoryless device. For example, the characteristic of an arbitrary phase detector could be measured and the resulting characteristic then represented by a Fourier series. The accuracy of the resulting approximation is arbitrary and the approximation can be made as precise as desired as long as the number of terms used in the Fourier series approximation is not constrained. There is, obviously, a trade off between the number of terms used in the Fourier series approximation (accuracy) and the resulting simulation run time although the penalty associated with including additional terms is very small.

The sawtooth-wave phase detector can be implemented in MATLAB using the piecewise-linear expression

$$s2 = rem(s1 + sign(s1) * pi, 2 * pi) - sign(s1) * pi \tag{4.6}$$

Note that (4.6), unlike a Fourier series representation, is exact since the sawtooth wave can be represented exactly using piecewise-linear line segments. Using (4.6) for the phase detector model results in the simulation program given in the m-file **pll2saw.m**.

The second-order phase-locked loop with transport delay is illustrated in Figure 4.4. Except for the addition of the transport delay block, it is identical to Figure 4.2. The transport delay is implemented as a shift register in which samples are shifted sequentially through the shift register at each simulation step. This is easily accomplished by establishing a vector, **sreg**, in which the input is placed in the first element of the vector and the output is taken as the last element in the vector. It therefore follows that the length of the vector is determined by the transport delay, as measured in sample periods, and is one plus the transport delay. Assuming that **s6** is the input to the shift register and **s7** is the output, as shown in Figure 4.4, we have the MATLAB code

$$sreg = [\ s6, \quad sreg(1,1:ndelp1-1) \ ] \tag{4.7}$$

and

$$s7 = sreg(1,ndelp1) \tag{4.8}$$

In the above expressions, **ndelp1** is the length of the vector representing the shift register. Equation (4.7) places the input in the first element of the vector **sreg** and shifts the remaining elements to the right by one index. Equation (4.8) extracts the last element of the vector and assigns it to the output. Equations (4.7) and (4.8) yield the m-file **pll2tdel.m**. Note that **pll2tdel.m** assumes a sinusoidal-wave phase detector. PLL models with transport delay and a triangular-wave phase detector or with transport delay and a sawtooth-wave phase detector are easily developed by simply inserting (4.7) and (4.8) into the appropriate code[3].

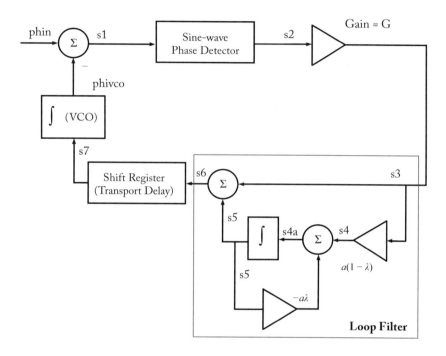

Figure 4.4: Second-order PLL Simulation Model with Transport Delay

The third-order phase-locked loop is illustrated in Figure 4.5 and the code is given in the m-file **pll3sin.m**. A sinusoidal-wave phase detector is assumed although this is easily changed if desired. A third-order system cannot be parameterized in terms of simple parameters such as the loop order and the damping factor and so the loop gain and the filter parameters are entered directly. Also note that the third-order loop is assumed "perfect".

---

[3]It would be very easy to develop a single m-file combining all loop filters, phase detectors, and including transport delay. This was not done in order to keep the code short, uncomplicated and therefore easily readable.

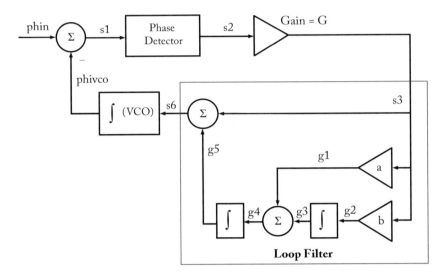

Figure 4.5: Third-order PLL Simulation Model

4.2   THE SIMULATION MODEL FOR THE COSTAS PLL

Prior to looking at the Costas PLL BPSK receiver we need to first examine the input signal. The assumed input signal is

$$x_{in}(t) = m(t)\cos[2\pi f_c t + \phi(t)] \tag{4.9}$$

in which the amplitude scaling of the input is normalized to unity. As discussed in Appendix A, a complex envelope signal is typically used when the input bandpass signal has both amplitude and phase variations. We there represent the input signal to the simulation model of the Costas PLL receiver by

$$\tilde{x}_{in}(t) = m(t)e^{j\phi(t)} \tag{4.10}$$

In (4.9) and (4.10), $m(t)$ represents the BPSK waveform and is either +1 or -1 in each symbol period.

The Costas PLL receiver is shown in Figure 4.6 and the simulation code is contained in **cpll.m**. The output of the VCO is[4]

$$\textbf{phivco}(t) = \theta(t) \tag{4.11}$$

The block labeled "make complex" yields output

$$\textbf{s1}(t) = e^{-j\theta(t)} \tag{4.12}$$

[4]Once again note that all variable names used in the MATLAB code are shown in boldface.

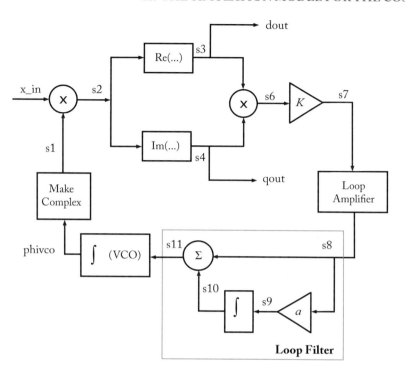

Figure 4.6: Simulation Model for the Costas Phase-Locked Loop

The loop input is, from (4.10),

$$\mathbf{xin}(t) = m(t)e^{j\phi(t)} \tag{4.13}$$

Multiplying (4.12) and (4.13) yields

$$\mathbf{s2}(t) = m(t)e^{j[\phi(t)-\theta(t)]} \tag{4.14}$$

The real and imaginary parts of **s2** are next formed. This gives

$$\mathbf{s3}(t) = m(t)\cos[\phi(t) - \theta(t)] \tag{4.15}$$

for the direct-channel (real) output, **dout**, and

$$\mathbf{s4}(t) = m(t)\sin[\phi(t) - \theta(t)] \tag{4.16}$$

for the quadrature-channel (imaginary) output, **qout**. For the case in which the input amplitude is normalized to one, $\mathbf{s3}(t)$ and $\mathbf{s4}(t)$ are exactly the same as $y_d(t)$ and $y_q(t)$ as described by (3.2) and (3.3), respectively. The remaining signals around the loop are as defined in Section 3.1.

Some words are in order concerning the amplifier having gain K. This gain is referred to as *excess gain*, and its inclusion in the model, although not strictly necessary, is motivated by the following argument. Multiplying **s3**(t) and **s4**(t) yields

$$\mathbf{s6}(t) = \frac{1}{2}m^2(t)\sin[2(\phi(t) - \theta(t))] \tag{4.17}$$

Assuming that $m^2(t) = 1$, as we assumed in Chapter 3 for BPSK demodulation, and $K = 1$ we have

$$\mathbf{s7}(t) = \frac{1}{2}\sin[2(\phi(t) - \theta(t))] \tag{4.18}$$

At this point, we find ourselves face with a dilemma. For operation where the phase error is small, in other words, for operation near phase lock (the tracking mode), the input to the loop filter is $\phi(t) - \theta(t)$ as we desire. In the acquisition mode, however, the phase errors are often large and, as shown in (4.18), the maximum value of the signal is 1/2 rather than 1 as in the phase-lock loop model. We can restore this value by letting $K = 2$. A reduction in gain of course increases the acquisition time. The dilemma arises because we wish to compare the performance of the Costas PLL with the performance of the perfect second-order PLL having the same loop natural frequency and loop damping factor. The loops parameters are based on the linear model, and it is the nonlinearity (the sinusoidal phase detector model in this case) that causes the difficulty. Note that the excess gain, K, is in series with the loop gain, G. Since this portion of the loop is linear, a single gain, having value KG, could be used.

The strategy taken here is to define the loop natural frequency and damping factor based on the perfect second-order PLL parameters as defined by (2.52) and (2.53). This, of course, gives the loop gain and natural frequency for the Costas PLL with $K = 1$ in Figure 4.6. If desired, we can then define the excess gain in the preprocessor to be different from this value. This is equivalent to changing the loop gain, G, with the filter parameter, a, held constant. This will, of course, change the loop natural frequency and the damping factor. Equations (2.52) and (2.53), together with a little algebra, show that non-unity values of K result in modified values of loop natural frequency and the damping factor, f_n^* and ζ^*, respectively, given by

$$f_n^* = \sqrt{K} f_n \tag{4.19}$$

and

$$\zeta^* = \sqrt{K} \zeta \tag{4.20}$$

where f_n and ζ are the equivalent perfect second-order PLL values of the loop natural frequency and damping factor.

4.3   THE QPSK LOOP

The simulation code for the QPSK Loop is contained in the file **qcpll.m**. The corresponding block diagram is illustrated in Figure 4.7. The assumed input to the loop is

$$\mathbf{xin}(t) = \tilde{m}(t)e^{j\phi(t)} \tag{4.21}$$

which is identical to the input for the Costas PLL discussed in the previous section except that the real message signal $m(t)$ has now been replaced by the complex message signal $\tilde{m}(t)$. The complex message, $\tilde{m}(t)$, is defined as

$$\tilde{m}(t) = m_d(t) + jm_q(t) \tag{4.22}$$

where $m_d(t)$ represents the direct-channel data and $m_q(t)$ represents the quadrature-channel data.

The complex version of the VCO output is defined by

$$\mathbf{s1}(t) = e^{-j\theta(t)} \tag{4.23}$$

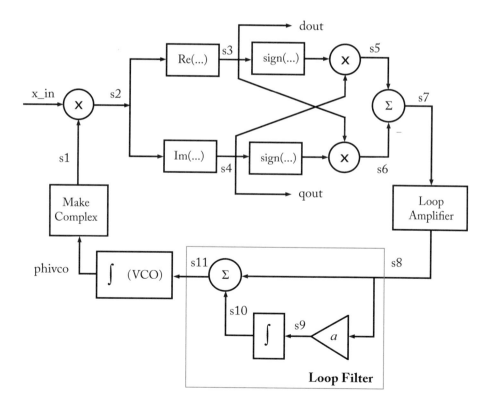

Figure 4.7: Simulation Model for the QPSK Loop

so that **s2**(t) can be written

$$\mathbf{s2}(t) = \tilde{m}(t)e^{j[\phi(t)-\theta(t)]} \tag{4.24}$$

which, expect for the fact that the message signal is complex, is identical to the equivalent signal in the Costas PLL. Taking the real and imaginary parts yields

$$\mathbf{s3}(t) = m_d(t)\cos[\phi(t) - \theta(t)] - m_q(t)\sin[\phi(t) - \theta(t)] \tag{4.25}$$

and

$$\mathbf{s4}(t) = m_d(t)\sin[\phi(t) - \theta(t)] + m_q(t)\cos[\phi(t) - \theta(t)] \tag{4.26}$$

respectively. It clearly follows from (4.25) and (4.26) that, for the case of *zero phase error*

$$\mathbf{s3}(t) = m_d(t) \tag{4.27}$$

and

$$\mathbf{s4}(t) = m_q(t) \tag{4.28}$$

as desired. If the phase error is non-zero, the loop dynamics cause the phase error to be driven towards zero. For sufficiently small phase error, (4.27) and (4.28) then become valid.

4.4   THE *N*-PHASE TRACKING LOOP

The block diagram of the *N*-Phase tracking Loop is shown in Figure 4.8. The corresponding code is contained in the file **npll.m** and is based on Figure 3.6. Note that there are N VCO outputs, N multipliers and N inputs to the product device. Thus, this part of the system is best simulated using vectors with each vector having N components.

The vector **phia** is a vector of N constants representing the N phase shifts of the VCO output. Thus from Figure 3.6 we see that

$$\mathbf{phia} = [\; 0 \quad \frac{\pi}{N} \quad \frac{2\pi}{N} \quad \cdots \quad \frac{(N-1)\pi}{N} \;] \tag{4.29}$$

The constant **pvco** is

$$\mathbf{pvco} = [\; \theta \quad \theta - \frac{\pi}{N} \quad \theta - \frac{2\pi}{N} \quad \cdots \quad \theta - \frac{(N-1)\pi}{N} \;] \tag{4.30}$$

Since the input to the lowpass model is the phase of the bandpass signal, it follows that the vector **s1** is given by

$$\mathbf{s1} = \exp(j[\; \phi - \theta \quad \phi - \theta + \frac{\pi}{N} \quad \phi - \theta + \frac{2\pi}{N} \quad \cdots \quad \phi - \theta + \frac{(N-1)\pi}{N} \;]) \tag{4.31}$$

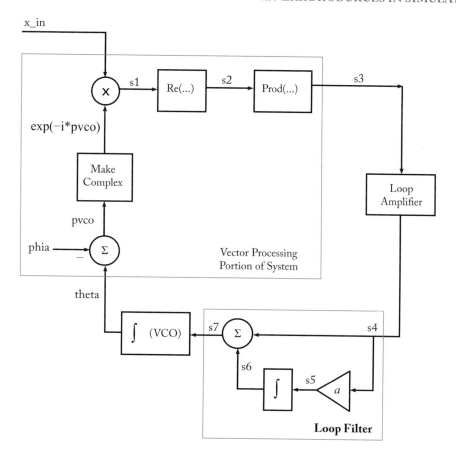

Figure 4.8: Simulation Model for the N-Phase Tracking Loop

The real part of each of the N vector components of **s1** is then taken. This gives the vector

$$\textbf{s2} = \cos([\ \phi - \theta \quad \phi - \theta + \tfrac{\pi}{N} \quad \phi - \theta + \tfrac{2\pi}{N} \quad \cdots \quad \phi - \theta + \tfrac{(N-1)\pi}{N}\ ]) \tag{4.32}$$

so that the jth component of (4.32) is

$$\textbf{s2}(j) = \cos(\phi - \theta + \frac{(j-1)\pi}{N}) \tag{4.33}$$

as shown in Figure 4.8. The rest of the simulation proceeds in a straightforward manner.

4.5   ERROR SOURCES IN SIMULATION

There are many potential sources of error when one develops a simulation of any system. To cover all of them is well beyond the scope of this book, and one should consult a textbook devoted to the subject. Here we simply briefly review some of the basic concepts.

The models presented in the first four sections of the chapter are all *simulation models*. Simulation models (18) are the end product of a process that begins with the physical device to be modeled, which usually involves hardware, such as single-chip implementation of a phase-locked loop. The process is shown in Figure 4.9. The first step in the modeling process is to develop an analytical model that captures the *essential features* of the physical device. The identification of these essential features requires engineering judgment and a thorough understanding to the application for which the models are being developed.

Figure 4.9: Process of Developing a Simulation Model

The first step in the derivation of an analytical model is to carefully specify those characteristics of the device being modeled that are to be represented in the analytical model. The resulting analytical model most often takes the form of a set of equations that describe the important input-output relationships of the physical device. These equations, at best, only approximate the device being modeled since they specify only certain aspects of the device behavior and are only accurate over a certain range of voltages, currents and frequencies. In summary, the analytical model is only an approximation of certain aspects of the physical device, and an understanding of these approximations is essential.

For a simple and brief example, we consider the physical device to be a resistor. The typical analytical model in the audio frequency range might be (usually is) the expression

$$e = Ri \qquad (4.34)$$

where e represents the voltage across the resistor, i represents the current through the resistor and R is a parameter known as resistance. The parameter R is usually considered to be a constant. Note that this simple model contains a very large number of assumptions. We consider only a very few here. It clearly follows that the model developed for the audio frequency range is not the same model needed at microwave frequencies. At microwave frequencies, the model may also require inductance and capacitance to be considered, among other things. Also note that, even at low audio frequencies, the resistance R will not be a constant if the model is to be applied to environments having extreme temperature variations. Also, if the model is to be used for determining a printed circuit board layout, the physical dimensions of the resistor will be important.

As an example closer to home, consider the problem of modeling a voltage-controlled oscillator (VCO). An ideal VCO is an oscillator in which the output frequency deviation is proportional to the input signal level. Letting $\theta(t)$ be the VCO phase deviation and $e(t)$ be the VCO input single level we have the VCO model used in Chapter 2. This was

$$\frac{1}{2\pi}\frac{d\theta}{dt} = K_d e(t) \tag{4.35}$$

where K_d represents the VCO constant in Hertz per unit of input signal. While this expression certainly captures the essential characteristic of a VCO, it may not be sufficient for many applications. For example, practical VCOs have associated with them a certain bandwidth which limits how quickly the VCO output can change from one frequency to another. If the bandwidth of the VCO is significantly greater than the open-loop bandwidth of the PLL, the VCO bandwidth can be considered infinite and (4.35) may apply.

Another modeling error might arise from the finite "rail voltage" of the VCO. This will constrain the maximum frequency deviation of the VCO. For example, if the hardware implementation limits $e(t)$ to a range of ± 5 volts, then the maximum frequency peak-to-peak deviation of the VCO is $10K_d$ Hertz. The analytical model, as expressed by the (4.35) does not model this effect and therefore (4.35) will only accurately model the physical device if $e(t)$ remains in the range $(-5,5)$ throughout the simulation. If the possibility of $e(t)$ exceeding this range in the simulation is significant, several options exist. An obvious solution would be to develop a more complex model that takes the finite rail voltage into account. This approach does, however, have drawbacks in that the more complex model will most likely require measurement data of the particular VCO to be modeled. Engineering time will be required to make the measurements and to develop the model. In addition, the more complex model is likely to significantly increase the simulation run time. A reasonable approach for some applications might be to simply use (4.35) as the analytical model, base the simulation on this model, and alert the user of the simulation user if $e(t)$ moves outside of the $(-5,5)$ range. In this way, the user can be aware of the fact that an incorrect (or at least incomplete) model is being used and that the simulation results must be viewed with this in mind.

Additional approximations and assumptions are involved in moving from an analytical model to a simulation model. We once again consider only a few. The voltages and currents in the physical device and the signals considered in the analytical model are typically continuous functions of the continuous variable time. In a discrete-time or digital simulation, these continuous-time signals are typically sampled and quantized to form discrete-time signals. The process of sampling and quantizing leads to aliasing and quantizing errors, respectively. While quantizing errors are typically negligible in simulations performed on floating-point processors, aliasing errors require our attention. Aliasing errors are reduced by increasing the sampling frequency, but this in turn means that more samples are required to define a time segment of data having fixed duration. It follows that increasing the sampling frequency also increases the time required to execute the simulation (run time). A trade off therefore exists between simulation run time and aliasing errors. One should not attempt to completely eliminate aliasing errors but rather to develop a simulation that achieves *reasonable* simulation run times with *acceptable* levels of aliasing error.

An important source of error resulting from the simulation strategy taken here results from the fact that a one sample period delay occurs in the simulation loop. This is illustrated in Figure 4.10 in which the desired continuous-time signals (CTS) are compared to the desired and actual discrete-time signals (DTS) that are processed by the simulation. Feedback loops give rise to interesting complications in simulations as illustrated in Figure 4.10. In the simulation of the phase-locked loop, the input is combined with the VCO output to form the phase detector output. The phase

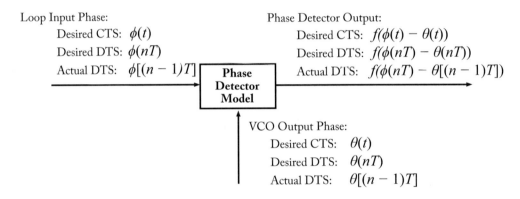

Figure 4.10: Illustration of a One Sample Period Transport Delay

detector output is then used to form the next VCO output. Consider for example the nth simulation step. The input to the phase detector is $\phi(nT)$, where T is the simulation step size. The output of the VCO was computed at the previous, $(n-1)^{th}$, simulation step. Thus instead of the phase detector output being defined by

$$\mathbf{s2}(t) = f\left[\phi(nT) - \theta(nT)\right] \tag{4.36}$$

as it should be to reflect proper operation of the hardware, it will be calculated as

$$\mathbf{s2}(t) = f\left[\phi(nT) - \theta((n-1)T)\right] \tag{4.37}$$

In the preceding two equations, the function $f(\dots)$ depends upon the phase detector characteristic and will be different for sine-wave, triangular-wave and sawtooth-wave phase detectors. Comparing (4.36) and (4.37), we see that an undesired transport delay of one sample period has been injected into the loop transfer function. *Is this significant?* The answer to this question depends, of course, upon the necessary accuracy of the simulation, which is in turn a function of the application. The effect of a one sample period transport delay clearly becomes less significant as the sampling frequency increases. In general, we can say that if the sampling frequency is small compared to the open-loop transfer function, the effect of the one sample period delay can *usually* be neglected.

Other approximations besides those mentioned here are involved in moving from an analytical model to a simulation model. We have seen that continuous-time integration, if present in the analytical model, must be replaced by discrete-time integration in the simulation model. There are many ways of accomplishing this and the trapezoidal approximation used in the MATLAB models developed in this book is simply one of the most basic (18). We could have taken the approach of representing the loop filter transfer function as a ratio of polynomials in the Laplace variable s and then mapping the transfer function in s to a ratio of polynomials in z^{-1} using impulse invariance, step invariance, the bilinear z-transformation or some other techniques for mapping a function of s into a function of z^{-1}. These techniques are all equivalent to specifying an integration algorithm and, as a matter of fact, choosing the bilinear z-transformation transformation is equivalent to replacing continuous-time integrators by discrete-time trapezoidal integration. All of these techniques are subject to different types of errors such as aliasing errors and frequency warping. Most of these sources of error also become small as the sampling frequency increases, but, as we have seen, this comes at the expense of increased simulation run time.

4.6   PROBLEMS

4.1 Show that the transfer function **s6/s3** of the loop filter simulation model given in Figure 4.2 is the same as the transfer function given in (2.20).

4.2 Using a MATLAB program, show that equation (4.2) yields the triangular waveform and equation (4.6) yields the sawtooth waveform as shown in Figure 4.3.

4.3 Using the triangular-wave phase detector in **pll2tdel.m** with the PLL parameters given in Numerical Example 2.5, compare its result with the one shown in Figure 2.15, which is obtained by using the sinusoidal phase detector in the simulation model.

4.4 Using the sawtooth-wave phase detector in **pll2tdel.m** with the PLL parameters given in Numerical Example 2.5, compare its result with the one shown in Figure 2.15, which is obtained by using the sinusoidal phase detector in the simulation model.

4.5 Show that the transfer function **s6/s3** of the loop filter simulation model given in Figure 4.5 is the same as the transfer function given in (2.22).

4.6 Verify equations (4.25) and (4.26).

4.7 For the N-phase tracking loop, find **s2** for $\mathbf{xin}(t) = [m_d(t) + jm_q(t)]e^{j\phi(t)}$. Note that **s2** of (4.32) is obtained from $\mathbf{xin}(t) = e^{j\phi(t)}$.

4.8 In Problem 4.7, let $N = 4$, find $\mathbf{s2}(2)$ and $\mathbf{s2}(4)$. Let $y_d(t) = \mathbf{s2}(2)$ and $y_q(t) = \mathbf{s2}(4)$, complete Table 4.1 below for $y_d(t)$ and $y_q(t)$ in terms of either $m_d(t)$ or $m_q(t)$ (similar to Table 3.2).

Table 4.1: Steady-state output of the 4-phase tracking loop

Steady-State Phase Error Ψ_{ss}	Output $y_d(t)$	Output $y_q(t)$
$\pi/4$		
$3\pi/4$		
$5\pi/4$		
$7\pi/4$		

CHAPTER 5

MATLAB Simulations

The purpose of this chapter is to demonstrate several of the simulation programs developed in the previous chapter. Not all, or even most, of the simulation programs are discussed here as the user of the simulations is encouraged to experiment, make changes and in general expand upon the basic capabilities illustrated here. Specifically, the second-order PLL and the QPSK demodulation loop are discussed in some details in the following pages. First, however, we pause to examine the structure of the simulation programs and the assumed loop inputs.

5.1   SIMULATION STRUCTURE

All simulations discussed in this chapter follow the structure illustrated in Figure 5.1. Three different operations are depicted in Figure 5.1: the preprocessor, the simulator and the postprocessor. These three operations are contained in all simulation programs although they may be merged to some extent in commercial simulation packages. It is strongly felt that partitioning a simulation program as depicted in Figure 5.1 will result in code that is more easily understood and maintained. As a bonus, the preprocessor and postprocessor quite often contain blocks of code that can be carried from one simulation project to another. Thus, for each example simulation to follow, there are three MATLAB script files: one for the preprocessor, one for the simulator and one for the postprocessor.

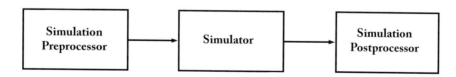

Figure 5.1: Simulation Structure

The purpose of the *preprocessor* is to specify all intrinsic simulation parameters and all system parameters. The intrinsic simulation parameters relate to those parameters that control the operation of the simulation without regard to the particular system being simulated. Examples of the intrinsic simulation parameters are the sampling frequency (or equivalently the simulation step size) and the length of the simulation run (in samples). Examples of system parameters include such things as filter bandwidths, amplifier gains, and signal parameters. In a MATLAB program, one may also wish to specify a number of other constants and vectors to be used by the simulator or the postprocessor. An example might be a vector of the sample times to be used in the postprocessor for plotting the time axis. The preprocessor is usually interactive so that the user can input various parameters as required to specify the simulation. Once all of the necessary parameters and other information required for execution of the simulation are specified, the resulting data is written to the MATLAB workspace.

The *simulator* (sometimes referred to as the simulation exerciser) reads the quantities stored on the MATLAB workspace and performs the desired simulation. The simulator should completely perform its task upon entering the name of the m-file containing the MATLAB script for the simulation. No intervention by the user should be necessary. A simulation is, of course, executed in order to generate certain quantities for later investigation. These may be numbers, such as a bit error rate or signal-to-noise ratio at a particular node in a system, or vectors defining a waveform at a point in a system. The simulation program writes all such information to the MATLAB workspace for later viewing by the postprocessor.

The *postprocessor* takes the information stored in the MATLAB workspace and operates upon it in a way that creates graphical displays and other information desired by the simulation user. All of the postprocessor outputs created by the programs to follow are activated using a simple menu. Although the desire to use MATLAB's "handle graphics" to implement radio buttons and slider bars was great, the temptation was resisted in order to keep the menu structure extremely simple so that the user of this set of programs can easily modify the postprocessor to perform other desired tasks.

The postprocessor, as implemented here, is more of a convenience than anything else. All functions performed by the postprocessor can, of course, be implemented by writing a short section of MATLAB code to generate the desired output. It is easy to generate graphical desired by the user in this way in the event that such desired displays are not included in the postprocessor menu. MATLAB has a rich collection of graphics functions and many useful displays can be generated by using them. Appropriately derived graphical displays can greatly facilitate the task of understanding complicated systems. Experimentation is therefore encouraged.

5.2   ASSUMED LOOP INPUTS

In general, the input to the PLL consists of two components: a frequency step and a frequency ramp. The frequency step is expressed in Hertz and the frequency ramp is expressed in Hertz per second. In other words,

$$\frac{1}{2\pi}\frac{d\phi}{dt} = Rt + \Delta_f \tag{5.1}$$

The loop input phase deviation, expressed in radians, is therefore

$$\phi(t) = \pi Rt^2 + 2\pi \Delta_f t + \theta_o \tag{5.2}$$

The simulation start time is assumed to be zero and the stop time is assumed to be t_f. The simulation is given a settle time equal to 10 percent of the total simulation run time, which in this case is t_s.

Thus the step and the ramp are applied at

$$t_s = t_f/10 \qquad (5.3)$$

These quantities are shown in Figure 5.2. In the preprocessor for the phase-locked loop simulations, the loop input consists only of a frequency step and R is assigned a default value of zero except for the case of the third-order PLL. For the third-order PLL, the value of Δ_f is set to the default value of zero and the user specifies the value of R. If desired, the preprocessor code can be easily modified so that the user can specify both Δ_f and R for any PLL configuration.

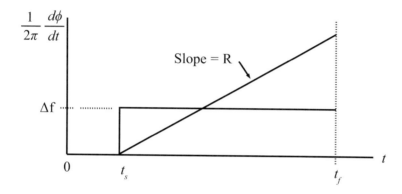

Figure 5.2: Simulation Structure

5.3   MATLAB AND SIMULINK SIMULATIONS

A number of MATLAB and SIMULINK simulations were developed and can be found at `http://www.morganclaypool.com/page/pll`. These are summarized in Tables 5.1 and 5.2, which show the file names for the preprocessor, simulator and postprocessor for each of the simulations. Note that all of the basic PLL simulations make use of the same preprocessor and postprocessor. The user is able to choose the loop order (1, 2 or 3) and is also able to choose between a perfect or imperfect loop within the preprocessor. The phase detector model is determined by the filename used for the simulation for the second-order PLL simulations. The Costas PLL and the QPSK loop have their own dedicated preprocessors and postprocessors. Note that since the N-Phase tracking loop is implemented for $N = 4$, it makes use of the preprocessor and postprocessor for the QPSK tracking loop.

Three SIMULINK files are also included as shown in Table 5.2. Examples of these SIMULINK simulations are contained in Appendix C.

Table 5.1: MATLAB Simulation File Names			
Description	**Preprocessor**	**Simulator**	**Postprocessor**
First-order Phase-locked Loop: sine-wave phase detector	pllpre.m	pll1sin.m	pllpost.m

Second-order Phase-locked Loop:
  sine-wave phase detector
  triangular-wave phase detector
  sawtooth-wave phase detector
  sine-wave phase detector
  with transport delay
	pllpre.m
	pllpre.m
	pllpre.m
	pllpre.m
		pll2sin.m
		pll2tri.m
		pll2saw.m
		pll2tdel.m
			pllpost.m
			pllpost.m
			pllpost.m
			pllpost.m

Third-order Phase-locked Loop: sine-wave phase detector	pllpre.m	pll3sin.m	pllpost.m
Costas Phase-locked Loop	cpllpre.m	cpll.m	cpllpost.m
QPSK Loop	qcpllpre.m	qcpll.m	qcpllpst.m
N-Phase Tracking Loop	qcpllpre.m	npll.m	qcpllpst.m

Table 5.2: SIMULINK File Names	
Description	**SIMULINK File**
Perfect Second-order PLL with Sine-wave Phase Detector	simpll2.mdl
Perfect Second-order PLL with Transport Delay	simpll2tdel.mdl
Perfect Third-order PLL with Sine-wave Phase Detector	simpll3.mdl

5.4 SECOND-ORDER PLL DEMONSTRATIONS

In the remainder of this chapter, we demonstrate several of the simulations included in Table 5.1. First, we consider the second-order PLL. We wish to use the same parameters given in Section 2.3.3 as follows:

$$\text{input frequency step} = 40 \text{ Hertz}$$

loop natural frequency = 10 Hertz
loop damping factor = 0.707
pole offset $\lambda = 0.2$

The first step in the simulation process is to invoke the preprocessor by entering **pllpre** at the MATLAB prompt. The parameter values shown above are then entered. The preprocessor dialog is shown in Figure 5.3. When all parameters have been entered, and all necessary calculations have been made, a MATLAB prompt appears. At this point, all necessary parameters required by the simulation program, as well as the frequency and phase deviations at the PLL input, are in the MATLAB workspace.

```
» pllpre

Enter order of loop (1, 2 or 3) > 2

Enter the size of the frequency step in Hertz > 40

Enter the loop natural frequency in Hertz > 10

Enter zeta, the loop damping factor > 0.707

Enter p for perfect or i for imperfect loop > i

Enter lambda > 0.2

» pll2tri
» pllpost
```

Figure 5.3: MATLAB Command Line Dialog for Second-Order PLL Simulation

It is important to note that a value of transport delay is entered in the preprocessor without regard to the simulation model that is eventually to be executed. The value of transport delay entered in the preprocessor is ignored unless the PLL model **pll2tdel** is executed.

In order to contrast this simulation with the simulation result given in Section 2.3.3, we execute the simulation using a triangular-wave phase detector. This is accomplished by entering **pll2tri** at the MATLAB prompt as shown. The simulation is then executed and the results are returned to the MATLAB workspace. At this point, one is ready to enter the postprocessor.

The postprocessor is entered by entering **pllpost** at the MATLAB prompt. Upon entering the postprocessor, a menu is displayed which lists the postprocessor options. This menu is illustrated in Figure 5.4.

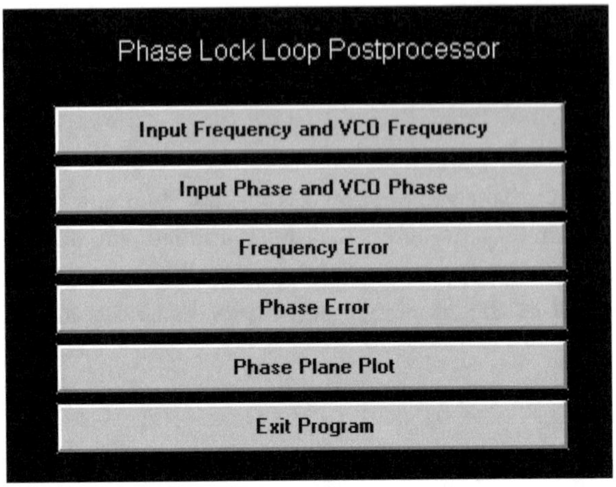

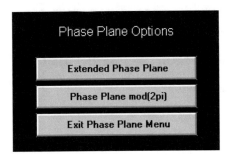

Figure 5.4: Postprocessor Menu for PLL Simulations

Figure 5.5: Submenu for Phase Plane Plots

With one exception, all of the items listed on the menu bars execute the calculations required to generate the requested display. The menu bar labeled "Phase Plane Plot" generates a submenu as illustrated in Figure 5.5. As shown in Figure 5.5, there are two different formats provided for phase-plane plots, the "Extended Phase Plane" and the "Phase Plane mod(2pi)". These two options are shown in Figures 5.6 and 5.7, respectively. The difference between the two options is clear; the phase plane, mod(2π), is constrained to lie between $-\pi$ and π while the extended phase plane does not impose this requirement. The extended phase plane is often more useful for examining the total number of cycles slipped since the final value of the steady-state error divided by 2π gives this important quantity. Requiring that the phase error lie between $-\pi$ and π yields a more conventional phase plane, such as those illustrated in textbooks. Note the non-zero, steady-state, phase error that results from the imperfect integration.

As we see in Figure 5.4, a number of displays, other than the phase plane, can be generated using the postprocessor. In order to conclude this section, one additional postprocessor output is illustrated. This is the display corresponding to the first menu bar, "Input Frequency and VCO Frequency". Clicking on this menu bar results in the display illustrated in Figure 5.8. Figure 5.8 shows the input frequency step occurring 10 percent into the simulation run and the initial frequency error is 40 Hertz and the final frequency error is 0. Figure 5.8 also shows that two cycles are slipped in the acquisition process, which is consistent with Figures 5.6 and 5.7.

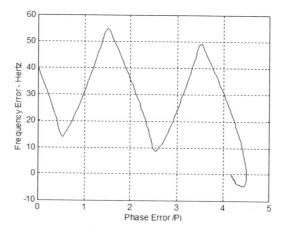

Figure 5.6: Extended Phase Plane Plot

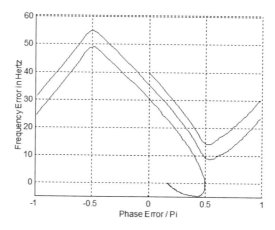

Figure 5.7: Phase Plane Plot mod(2pi)

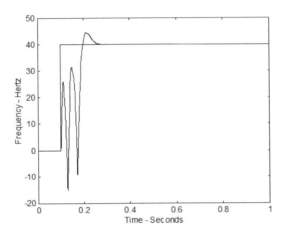

Figure 5.8: Input Frequency and VCO Frequency

5.5   QPSK LOOP

The QPSK loop is illustrated using the preprocessor dialog illustrated in Figure 5.9. The parameters chosen will cause the loop to slip cycles prior to achieving phase lock as we will see upon execution of the postprocessor. After the simulation parameters are entered in the preprocessor, the simulation is executed by entering **qcpll** as shown. Following execution of the simulation, the postprocessor is activated by entering **qcpllpst** at the MATLAB prompt. The postprocessor menu, illustrated in Figure 5.10, is then displayed.

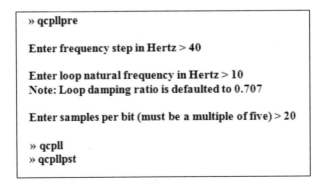

Figure 5.9: MATLAB Command Line Dialog for QPSK Loop Simulation

The extended phase plane is illustrated in Figure 5.11. Note that nine cycles are slipped before the loop achieves phase lock. We must, however, be careful to interpret this correctly. Note

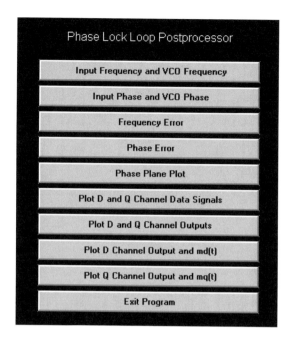

Figure 5.10: QPSK Postprocessor Menu

from Figure 5.11 that the loop appears to slip four cycles as the phase error varies between 0 and 2π radians. This corresponds to four slips per cycle of the input carrier. This results from the fact that the QPSK loop has four stable operating points per cycle of the input carrier as was discussed in detail in Chapter 3.

As with other simulations presented here, the frequency step that is applied 10 percent into the simulation run causes the tracking receiver to lose lock. We can also see from Figure 5.11 that, after reacquisition, the steady-state phase error ψ_{ss} is 4.5π. Thus ψ_{ss}, taken mod(2π), is $\pi/2$ radians. Using the analogy of input signal and demodulated[1] signal basis sets discussed in Chapter 3, the $\pi/2$ radian steady-state phase error will result in the direct channel signal appearing on the quadrature channel demodulator output. As a matter of fact, Table 3.2 predicts that the direct and quadrature channel demodulator outputs will be defined by $y_d(t) = -m_q(t)$ and $y_q(t) = m_d(t)$, respectively. Comparison of Figure 5.12, which illustrates the direct and quadrature channel data signals, with Figure 5.13, which shows the direct and quadrature channel demodulated outputs, shows this to be true.

[1]We refer to baseband signals as demodulated signals recognizing that these correspond to demodulated signals in the case of analog demodulation. Demodulation of digital signals (e.g., BPSK or QPSK) requires the addition of a digital demodulator such as an integrate-and-dump detector.

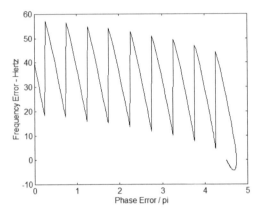

Figure 5.11: Phase Plane Plot for the QPSK Loop

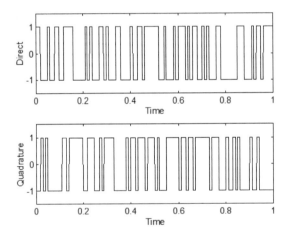

Figure 5.12: Direct and Quadrature Channel Data Signals

The loss of lock that results from the application of the frequency step can clearly be seen in Figure 5.13. The tracking receiver loses lock at $t = 0.1$ seconds, and it takes approximately 0.1 seconds for the receiver to reacquire phase lock.

5.6 THE N-PHASE TRACKING LOOP

The implementation of the simulation of the N-phase tracking loop is somewhat different from the other simulation presented in this chapter. The N-Phase tracking loop can be implemented for any value of N that is a multiple of four. For purposes of illustration, we use $N = 4$ here. This choice allows us to utilize the preprocessor (qcpllpre.m) and the postprocessor (qcpllpst.m) developed for

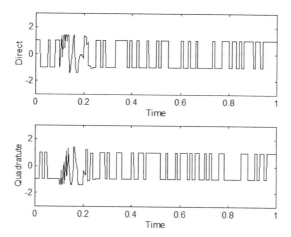

Figure 5.13: Direct and Quadrature Channel Demodulated Outputs

the QPSK loop for the *N*-Phase tracking loop.

There is one other significant difference between the simulation for the *N*-phase tracking loop and the preceding simulations. This difference results from the use of the postprocessor for the QPSK loop with the *N*-phase tracking loop. Since the *N*-phase tracking loop generates *N* output signals (*N* = 4 for the case under consideration) and the postprocessor expects two inputs, designated **dout** and **qout**, it is necessary to assign one of the *N* = 4 outputs of the N-phase tracking loop to **dout** and one to **qout** prior to execution of the postprocessor.

For the current demonstration of the *N*-phase tracking loop, the following two lines of MATLAB code are executed to make the necessary assignments:

$$>> \textbf{dout} = \textbf{out2};$$
$$>> \textbf{qout} = \textbf{out4};$$

These two lines of code are executed after execution of the simulation and prior to execution of the postprocessor. Observation of the MATLAB code for the *N*-phase tracking loop shows that the postprocessor inputs **dout** and **qout** can be assigned to any of the four outputs generated by the *N*-phase tracking loop, **out1**, **out2**, **out3** or **out4**. Note that since all four of these signals are written to the MATLAB workspace upon execution of the simulation of the *N*-phase tracking loop, reassignments of signals to **dout** and **qout** can be made without re-execution of the simulation.

Execution of the *N*-phase tracking loop simulation with the same parameters used for execution of the QPSK loop simulation, illustrated in Figure 5.9, produces the phase plane shown in Figure 5.14. The steady-state phase error is 8.75π which, taken $\mod(2\pi)$, is $3\pi/4$ radians.

The result of this phase error can be better understood by studying Figure 5.15. Note that for $N = 4$, the steady-state phase error is $k\pi/4$, where k is an integer. Assuming that the original data is the "usual" QPSK signal constellation shown in Figure 5.15(a), a steady-state phase error of $k\pi/4$ with k odd will yield the signal constellation shown in Figure 5.15(a), while k even will yield the signal constellation shown in Figure 5.15(b). It should also be noted that the signal constellation at the output of the receiver can be rotated $\pi/4$ radians by mapping **out2** and **out4** to **dout** and **qout**, as was done in this example, rather than mapping **out1** and **out3** to **dout** and **qout**, as might typically be done.

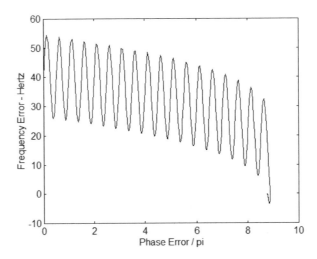

Figure 5.14: Phase Plane for N-Phase Tracking Loop Demonstration

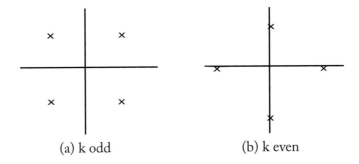

Figure 5.15: Possible QPSK Signal Constellations

It should be noted that the conventional QPSK signal constellation shown in Figure 5.15(a) leads to direct and quadrature channel signals that are two-level signals, while the QPSK signal constellation shown in Figure 5.15(b) leads to three-level direct and quadrature channel signals.

The direct and quadrature channel data signals resulting from execution of the simulation are shown in Figure 5.16. The baseband output signals are shown in Figure 5.17. A frequency step of 40 Hertz is applied at $t = 0.1$ seconds that results in the tracking receiver losing lock. The loss of lock can clearly be seen in Figure 5.17. It can also be observed from Figure 5.17 that it takes about 0.2 seconds for the loop to regain lock. Of particular interest is that prior to the application of the frequency step and the subsequent loss of lock, the direct and quadrature channel data signals correspond to the three-level signal constellation illustrated in Figure 5.15(a) while the baseband outputs correspond to the two-level signal constellation illustrated in Figure 5.15(b). This results because of the mapping of **out2** to **dout** and **out4** to **qout** that was performed prior to execution of the postprocessor.

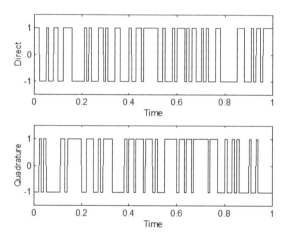

Figure 5.16: Direct and Quadrature Channel Data Signals

Since the steady-state phase error generated by the frequency step was an odd multiple of $\pi/4$ resulting in the signal constellation being transformed from the k even to the k odd constellations depicted in Figure 5.15, this mapping was necessary in order to generate conventional (two-level) direct and quadrature channel QPSK signals after reacquisition. This was desired so that comparisons could easily be made between the original data signals and the demodulated signals following reacquisition. The effect of the mapping was to rotate the signal space of the receiver $\pi/4$ radians with respect to the signal space of the original data signals.

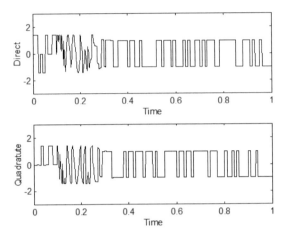

Figure 5.17: Direct and Quadrature Channel Baseband Outputs

Comparison of Figures 5.16 and 5.17 illustrates that, after reacquisition, the direct channel output is the inverted direct channel data signal. The quadrature channel output is the quadrature channel data signal.

5.7 PROBLEMS

5.1 The result given in Figure 5.7 was obtained by using a triangular-wave phase detector instead of a sinusoidal phase detector. Examine the impact of the phase detector by comparing the result with Figure 2.12.

5.2 Using the sawtooth-wave phase detector in the simulation of the imperfect second order PLL with the parameters given in Figure 5.3, compare its results with those given in Figures 2.12 and 5.7. Explain the impact of the phase detectors such as number of cycle slips in the results for each phase detector.

5.3 Using simulation, find the steady-state phase error $\mod(2\pi)$ for a second-order PLL with a sine-wave phase detector with $\lambda = 0.2$ (using the zoom-in feature on MATLAB figure window). Using the linear model (Table 2.1), find the steady-state phase error and compare it with the phase error from the simulation. Repeat for $\lambda = 0$ and $\lambda = 0.1$. Explain the impact of λ.

5.4 A phase detector is defined by the expressions

$$\begin{aligned}
y &= \sin(x), & 0 \leq x \leq \pi \\
y &= (2x/\pi), & -\pi/2 \leq x < 0 \\
y &= \sin(x), & -\pi \leq x < -\pi/2
\end{aligned}$$

Using the Fourier series technique described in Appendix A, execute the simulation described in Section 5.4 with the phase detector described by the preceding expression. Use a nine-term series approximation. Compare the results of the simulation with the sinusoidal phase detector used in Section 5.4.

5.5 Change the frequency step in Figure 5.9 to 50 Hz and keep the other parameters the same, find the steady-state phase error mod(2π) of the QPSK loop. Plot the direct and quadrature channel outputs and show that they are consistent with the outputs given in Table 3.2. How long does it take for the QPSK loop to regain phase lock?

5.6 Change the loop natural frequency in Figure 5.9 to 15 Hz and keep the other parameters the same, find the steady-state phase error mod(2π) of the QPSK loop. Plot the direct and quadrature channel outputs and show that they are consistent with the outputs given in Table 3.2. How long does it take for the QPSK loop to regain the phase lock?

5.7 The results of the QPSK simulation, illustrated in Figures 5.11–5.13, were based on the model illustrated in Figure 3.3, in which hard limiters were used at the lowpass filter outputs. A more standard nonlinearity is the hyperbolic tangent, tanh(x). Rerun the QPSK simulation presented in this chapter with the hyperbolic tangent substituted for the hard-limiter. Compare the results and discuss the differences if you observe any.

5.8 Run a simulation for the N-phase tracking loop to confirm the results given in Figures 5.14, 5.16, and 5.17.
Note that the relationship between the direct and quadrature channel data signals and the direct and quadrature channel outputs can be proved in Problem 4.8.

5.9 Run a simulation for the N-phase tracking loop with the frequency step of 36 Hz, the loop natural frequency of 10 Hz, and 20 samples per bit. Determine the steady-state phase error mod(2π) of the N-phase tracking loop. Using **dout** = **out2** and **qout** = **out4**, plot the direct and quadrature channel data signals and the direct and quadrature channel outputs. State the relationship between those two plots after reacquiring the phase lock.
Note that the relationship between the direct and quadrature channel data signals and the direct and quadrature channel outputs were shown in Problem 4.8.

CHAPTER 6

Noise Performance Analysis

In this chapter, we will study PLL in the presence of noise. No attempt is made to present a complete discussion since this is adequately accomplished in the literature. Our purpose is to provide basic understanding of PLL performance in the presence of noise. The PLL with additive Gaussian noise is analyzed. The discussion of nonlinear analysis is limited to the first-order PLL with additive noise. We also briefly talk about the effect of phase noise. The MATLAB simulation for the first-order PLL with additive noise is given in the last section.

6.1 PLL WITH ADDITIVE NOISE

Let an input to the sinusoidal phase detector of the basic PLL be the input signal plus noise as shown in Figure 6.1. Thus

$$x_{in}(t) = A_c \cos[2\pi f_c t + \phi(t)] + n(t) \tag{6.1}$$

where $n(t)$ is stationary and bandpass Gaussian noise. The signal at the output of the VCO is assumed to have the form

$$x_{vco}(t) = -A_v \sin[2\pi f_c t + \theta(t)] \tag{6.2}$$

The bandpass input noise can be written as (see Appendix A)

$$n(t) = n_d(t) \cos[2\pi f_c t] - n_q(t) \sin[2\pi f_c t] \tag{6.3}$$

where $n_d(t)$ and $n_q(t)$ represent stationary and statistically independent baseband Gaussian noise processes with zero mean and variance

$$\sigma_n^2 = E[n^2] = E[n_d^2] = E[n_q^2] \tag{6.4}$$

The one-sided spectra of $n_d(t)$ and $n_q(t)$ are

$$S_{nd}(f) = S_{nq}(f) = S_n(f_c - f) + S_n(f_c + f), \quad f \geq 0 \tag{6.5}$$

where $S_n(f)$ is the one-sided spectral density of the bandpass input noise $n(t)$. The noise spectral density is assumed to be nearly flat over the frequency range of the receiver so that the noise can be assumed to be white (7). The one-sided spectral density is $S_n(f) = N_0$ watts/Hz.

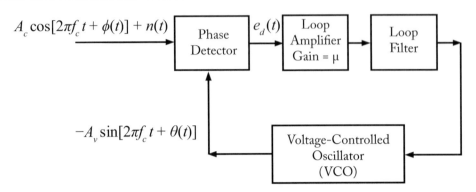

Figure 6.1: Basic Phase-Locked Loop Model

The output of the sinusoidal phase detector, consisting of a multiplier and a lowpass filter with a gain of 2, is therefore

$$e_d(t) = A_c A_v \sin[\phi(t) - \theta(t)] - A_v n_d(t)\sin[\theta(t)] + A_v n_q(t)\cos[\theta(t)] \tag{6.6}$$

The above equation can also be written as

$$e_d(t) = A_c A_v \left[\sin[\phi(t) - \theta(t)] + n'(t) \right] \tag{6.7}$$

where $n'(t)$ is an equivalent noise defined as

$$n'(t) = -\frac{n_d(t)}{A_c}\sin[\theta(t)] + \frac{n_q(t)}{A_c}\cos[\theta(t)] \tag{6.8}$$

Figure 6.2 shows a nonlinear phase model of the PLL with equivalent noise.

From the characteristics of the bandpass input noise (6.4) and assuming that $\theta(t)$ is independent of noise (1), it can be shown that $n'(t)$ has zero mean and its variance can be written as

$$\sigma_{n'}^2 = \frac{\sigma_n^2}{A_c^2} \tag{6.9}$$

Note that the variance does not depend on the value of $\theta(t)$.

For the case of white noise, it can be shown that $n'(t)$ can be treated as a white Gaussian process with a one-sided spectral density $S_{n'}(f) = 2N_0/A_c^2$ (7).

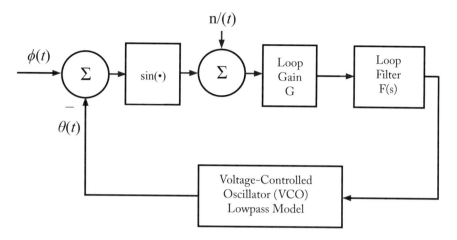

Figure 6.2: Nonlinear PLL Phase Model with Equivalent Noise

6.1.1   LINEAR ANALYSIS

When the phase error, $\phi(t) - \theta(t)$, is small, the sinusoidal phase detector can be considered to be operating in a linear mode. Figure 6.3 shows the linear PLL phase model with equivalent noise[1].

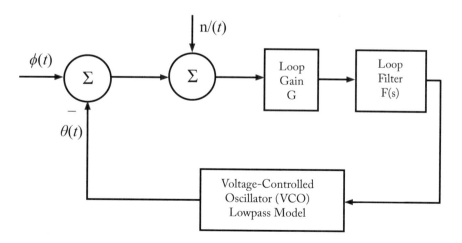

Figure 6.3: Linear PLL Phase Model with Equivalent Noise

---

[1]Since we now deal with a linear system, the superposition principle can be used. We can determine the effect of an input, and the effect of noise separately, and then combine them.

From Figure 6.3, we can see that the transfer function relating the VCO phase $\theta(t)$ to $n'(t)$ is the same as the transfer function, $H(s)$, relating the VCO phase $\theta(t)$ to the input phase $\phi(t)$, (2.13):

$$H(s) = \frac{\Theta(s)}{N'(s)} = \frac{G\frac{F(s)}{s}}{1 + G\frac{F(s)}{s}} \tag{6.10}$$

The spectral density of the VCO phase $S_{\theta n'}$ is related to the spectral density of the equivalent noise $S_{n'}$ as

$$S_{\theta n'}(f) = S_{n'}(f)|H(f)|^2 \tag{6.11}$$

where $H(f) = H(s)|_{s=j2\pi f}$.

6.1.2   NOISE BANDWIDTH

The variance of the VCO phase due to the noise is an integral of (6.11):

$$\sigma_{\theta_n}^2 = \int_0^\infty S_{n'}(f)|H(f)|^2 df \tag{6.12}$$

For the special case of white noise, $S_{n'}(f) = 2N_0/A_c^2$, (6.12) is simplified to

$$\sigma_{\theta_n}^2 = \frac{2N_0}{A_c^2} \int_0^\infty |H(f)|^2 df \tag{6.13}$$

The **noise bandwidth** B_L is defined as

$$B_L = \int_0^\infty |H(f)|^2 df \tag{6.14}$$

For the *first-order* PLL, $H(f) = \frac{G}{j2\pi f + G}$, the noise bandwidth is

$$B_L = \frac{G}{2\pi} \int_0^\infty \frac{1}{1 + x^2} dx = \frac{G}{4} \tag{6.15}$$

where $x = 2\pi f/G$.

For the *perfect-second order* PLL, $H(f) = \frac{G(j2\pi f + a)}{-(2\pi f)^2 + G(j2\pi f) + Ga}$, the noise bandwidth is

$$B_L = \frac{G}{2\pi} \int_0^\infty \frac{(\frac{a}{G})^2 + x^2}{x^4 + (1 - 2\frac{a}{G})x^2 + (\frac{a}{G})^2} dx = \frac{G}{4}\left(1 + \frac{a}{G}\right) \tag{6.16}$$

where $x = 2\pi f/G$.

Using a similar approach for the imperfect second order PLL and the third order PLL, their noise bandwidths can be determined. Table 6.1 summarizes the noise bandwidths of PLLs. Note that for second-order PLLs, G and a can be written in terms of the damping factor ζ and the natural frequency f_n as $G = 4\pi\zeta f_n$ (2.52) and $a = \pi f_n/\zeta$ (2.53).

Table 6.1: Noise Bandwidths of PLLs

Loop Type	Loop Filter $F(s)$	VCO Phase to Input Phase Transfer Function $\frac{\Theta(s)}{\Phi(s)} = H(s)$	Noise Bandwidth B_L
First order	1	$\dfrac{G}{s+G}$	$\dfrac{G}{4}$
Second order (Perfect)	$1 + \dfrac{a}{s}$	$\dfrac{G(s+a)}{s^2 + Gs + Ga}$	$\dfrac{G}{4}\left(1 + \dfrac{a}{G}\right)$
Second order (Imperfect)	$\dfrac{s+a}{s+\lambda a}$	$\dfrac{G(s+a)}{s^2 + (G+\lambda a)s + Ga}$	$\dfrac{G}{4}\left(\dfrac{G+a}{G+\lambda a}\right)$
Third order	$1 + \dfrac{a}{s} + \dfrac{b}{s^2}$	$\dfrac{G(s^2 + as + b)}{s^3 + Gs^2 + Gas + Gb}$	$\dfrac{\dfrac{G}{4}\left(1 + \dfrac{a}{G} + \dfrac{b}{Ga}\right)}{1 - \dfrac{b}{Ga}}$

From Table 6.1, several observations can be made:

- Loop gain G is important in establishing the noise bandwidth. The smaller the G value, the smaller the noise bandwidth. However, the larger the steady-state error becomes (Table 2.1). There is a trade off between the noise bandwidth and the steady-state error when choosing G.

- The imperfect second-order PLL: When $\lambda = 0$, the noise bandwidth is the same as that of the perfect second order PLL, and when $\lambda = 1$, the noise bandwidth is the same as that of the first order PLL.

- The perfect second-order PLL, B_L can also be written as

$$B_L = \pi f_n \left(\zeta + \frac{1}{4\zeta}\right)$$

The damping value, ζ, of 0.5 yields the minimum noise bandwidth of $B_L = \pi f_n = 0.5\omega_n$. The noise bandwidth does not exceed the minimum by more than 25 percent for any damping value between 0.25 and 1.0 (1).

6.1.3   SIGNAL TO NOISE RATIO OF THE LOOP

The noise output $n'(t)$ of the phase detector could be caused by the additive noise as described earlier or by a disturbance in the input phase $\sin(\phi_n(t)) = n'(t)$. If ϕ_n is small enough, the linear estimation, $\phi_n(t) \approx n'(t)$, can be used and the variance of this input phase disturbance is $\sigma_{\phi_n}^2 = \sigma_{n'}^2 = \sigma_n^2 / A_c^2$ (1).

Let the signal-to-noise ratio at the input of the PLL be defined as $(SNR)_i = \dfrac{A_c^2/2}{\sigma_n^2}$, the input phase variance[2] is

$$\sigma_{\phi_n}^2 = \frac{\sigma_n^2}{A_c^2} = \frac{1}{2(SNR)_i} \tag{6.17}$$

With the noise bandwidth definition (6.14), the variance of the VCO phase caused by the white noise input (6.13) is

$$\sigma_{\theta_n}^2 = \frac{2N_0}{A_c^2} B_L \tag{6.18}$$

By analogy, we can define a signal-to-noise ratio at the output (SNR of the loop) as

$$\sigma_{\theta_n}^2 = \frac{1}{2(SNR)_L} \tag{6.19}$$

From (6.18) and (6.19), the **SNR of the loop** is obtained as

$$(SNR)_L = \frac{A_c^2/2}{2N_0 B_L} \tag{6.20}$$

Note that this $(SNR)_L$ is applicable only for linear operation in a white noise environment.

6.1.4   NONLINEAR ANALYSIS

Linear analysis works well when $(SNR)_L$ is reasonably large since the phase error is small and the linear approximation is adequate. However, at low $(SNR)_L$, nonlinear analysis is required. The nonlinear analysis of PLL with noise requires a higher level of mathematical complexity than the linear analysis. In this section, we discuss only the nonlinear analysis of the first order PLL. For higher order PLLs, interested readers can refer to the references such as (6), (8), and (13).

---

[2]This variance is valid for large $(SNR)_i$ (1)

The nonlinear analysis of the first-order PLL with noise is derived by Viterbi (12), (7). Let the input to the loop and the output of the VCO be

$$x_{in}(t) = A_c\cos[2\pi f_c t + \phi(t)] + n_d(t)\cos[2\pi f_c t] - n_q(t)\sin[2\pi f_c t]) \tag{6.21}$$

and

$$x_{vco}(t) = -A_v\sin[2\pi f_c t + \theta(t)]. \tag{6.22}$$

The output of the phase detector, consisting of a multiplier and a lowpass filter with a gain of 2, is

$$e_d(t) = A_c A_v \left[\sin[\phi(t) - \theta(t)] - \frac{n_d(t)}{A_c}\sin\theta(t) + \frac{n_q(t)}{A_c}\cos\theta(t) \right] \tag{6.23}$$

After multiplication by the loop gain μ, the phase detector output is filtered by a loop filter with an impulse response $f(t)$. The input to the VCO is given by

$$e_{vco}(t) = \mu f(t) * e_d(t) \tag{6.24}$$

From equations (2.5) and (6.24), the output of the VCO is

$$\frac{d\theta}{dt} = G\,f(t) * \left[\sin[\phi(t) - \theta(t)] - \frac{n_d(t)}{A_c}\sin\theta(t) + \frac{n_q(t)}{A_c}\cos\theta(t) \right] \tag{6.25}$$

where $G = 2\pi K_d \mu A_c A_v$.

With the definition of $n'(t)$ and the phase error $\psi(t) = \phi(t) - \theta(t)$, we can write

$$\frac{d\psi(t)}{dt} = \frac{d\phi(t)}{dt} - G\,f(t) * \left[\sin[\psi(t)] + n'(t) \right] \tag{6.26}$$

If the input signal is assumed to have a constant phase ϕ and the loop is in lock, that is $f_{vco} = f_c$, (6.26) can then be simplified as

$$\frac{d\psi(t)}{dt} = -G\,f(t) * \left[\sin[\psi(t)] + n'(t) \right] \tag{6.27}$$

For the first order PLL, $f(t) = \delta(t)$, (6.27) then becomes

$$\frac{d\psi(t)}{dt} = -G\left[\sin[\psi(t)] + n'(t) \right] \tag{6.28}$$

Since $n'(t)$ is a white Gaussian process, the instantaneous change in ψ $(d\psi(t)/dt)$ depends only on the present value of ψ and the present value of the noise. This ψ is therefore a continuous Markov process, and **Fokker-Planck** techniques can be used to determine its probability density function (12).

In Viterbi's derivation, ψ is defined as ψ modulo 2π $(-\pi \leq \psi < \pi)$. It turns out that ψ is stationary in the steady-state. The steady-state probability density function[3] of ψ is given as (12)

$$p(\psi) = \frac{\exp(\rho \cos \psi)}{2\pi I_0(\rho)}, \quad -\pi \leq \psi < \pi \tag{6.29}$$

where $\rho = 2(SNR)_L$ and $I_0(\rho)$ is the modified Bessel function of the first kind and zero order.

When $(SNR)_L$ is large, the density (6.29) resembles Gaussian distribution with a variance of $1/\rho$, which is consistent with the result from the linear analysis. However, when $(SNR)_L$ is very small, the density approaches a uniform distribution $p(\psi) = \frac{1}{2\pi}$, $-\pi < \psi < \pi$ which yields a variance of $\pi^2/3$.

6.2   PLL WITH VCO PHASE NOISE

Every electronic component generates noise and the components of PLL are no exception. Noise sources such as thermal noise, shot noise, and flicker noise in the circuit will perturb the control voltage of VCO and result in variations of the VCO output frequency. The output spectrum will therefore contain other frequency components around the nominal VCO frequency, f_{VCO}[4]. The spread of the output spectrum which contains unwanted frequency components around the nominal VCO frequency is called *phase noise*. Unlike additive noise, phase noise is multiplicative, nonstationary, and has non-white spectrum. In this section, we discuss only basic analysis of PLL behavior with VCO phase noise[5].

The main source of phase noise in a PLL is the VCO. In general, the output of the VCO can be written as

$$x_{vco}(t) = (A_v + A_{vn}(t))\sin[2\pi f_{vco}t + \theta(t)]$$

where A_v and f_{vco} are the nominal VCO amplitude and frequency, respectively. $A_{vn}(t)$ is the amplitude noise and $\theta(t)$ contains all phase and frequency variations from f_{vco} which includes random phase noise, frequency offset and drift, and initial phase. Generally, the oscillator has an amplitude control mechanism that limits amplitude fluctuations, the effects of amplitude noise is therefore not considered.

Several definitions of the spectral density functions can be used to describe phase noise (1). Essentially, it measures the frequency or phase deviations from the nominal frequency of the oscillators. In this book, the baseband spectrum of the phase noise is used. The phase noise spectra

---

[3]This is the steady state solution of a nonlinear stochastic partial differential equation known as the Fokker-Planck equation.
[4]Note that if the VCO is ideal, its output spectrum as seen on a spectrum analyzer would be a single spectral line at f_{VCO}. In practice, this is not the case.
[5]The behavior of PLL in phase noise is still an open research topic. There is no solid theoretical basis for mathematical analysis of phase noise. A large amount of literature is available on this topic. The background articles and references on this topic can be found in (15).

of VCO, $S_{\theta_{nVCO}}(f)$, generally consists of discrete spectral lines such as periodic interference from AC power residues and the continuous spectral parts. The continuous phase noise spectra can be modeled by (14)

$$S_{\theta_{nVCO}}(f) = \frac{h_4}{f^4} + \frac{h_3}{f^3} + \frac{h_2}{f^2} + \frac{h_1}{f} + h_0$$

where h_i are constants specified for each noise source. The first term h_4/f^4 occurs mainly at frequencies below 1 Hz and generally causes no problem for VCO in PLL. The second term h_3/f^3 results from the flicker noise $(1/f)$ within the oscillator that causes the frequency fluctuations in the oscillator. The integration of these frequency fluctuations yields the phase fluctuations and produces the $1/f^3$ spectral components. The thermal (white) noise within the oscillator produces the third term, h_2/f^2 (1). The last two terms are also results of flicker noise and white noise, respectively.

6.2.1   LINEAR ANALYSIS OF VCO PHASE NOISE

The phase noise generated within the VCO can be modeled as shown in Figure 6.4, which consists of a noise-free ideal VCO with an output θ_{VCO} and a phase noise source θ_{nVCO}. The transfer function relating the phase noise θ_{nVCO} to the VCO output θ is given as

$$\frac{\Theta(s)}{\Theta_{nVCO}(s)} = \frac{s}{s + GF(s)} = E(s) \tag{6.30}$$

This is the same as the transfer function relating the phase error to the input phase, $E(s) = \frac{\Psi(s)}{\Phi(s)} = 1 - H(s)$, given in (2.16).

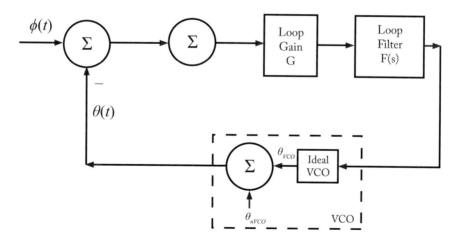

Figure 6.4: Model of Phase Noise in the Oscillator

From (6.30), the baseband spectrum of the phase noise $S_\theta(f)$ is

$$S_\theta(f) = S_{\theta_{nVCO}}(f)|E(f)|^2 \tag{6.31}$$

where $E(f) = E(s)|_{s=j2\pi f}$.

Note that the transfer function of (6.30) has a high pass frequency response. In order to reduce the phase noise, the high pass cutoff frequency of $E(f)$ must be increased since $S_\theta(f)$ is a function of $E(f)$. Recall that $E(s) = 1 - H(s)$ where $H(s)$, a PLL closed-loop transfer function, is a lowpass filter. Increasing the cutoff frequency of $E(f)$ implies increasing the cutoff frequency of $H(f)$ or increasing the PLL bandwidth. In Section 6.1, we learn that, to reduce the additive noise, the PLL bandwidth should be reduced since the additive noise is suppressed by $H(f)$, see (6.11). Hence, there is a trade off when designing the PLLs in order to minimize those two types of noise, in other words, the optimum cutoff frequency (bandwidth) should be determined in order to minimize the total phase variations caused by both noise.

The total variance of VCO phase caused by both additive noise (6.12) and phase noise (6.31) is

$$\sigma_\theta^2 = \int_0^\infty S_{n'}(f)|H(f)|^2 df + \int_0^\infty S_{\theta_{nVCO}}(f)|E(f)|^2 df \tag{6.32}$$

Optimization algorithms must be used in order to find the optimum parameters to minimize the total phase variance (6), (11). For example, in the second-order PLL which is defined in terms of ζ and f_n, with the assumption of white Gaussian noise and the approximation of $S_{\theta_{nVCO}}(f)$ in (6.32), we can find the optimum values of ζ and f_n such that the total variance of VCO phase is minimized.

6.3   SIMULATION OF 1ST-ORDER PLL WITH ADDITIVE NOISE

6.3.1   SIMULATION MODEL

The simulation model for the first-order PLL with additive noise is shown in Figure 6.5. This model[6] is based on the nonlinear PLL phase model with equivalent noise, Figure 6.2. To focus on the effect of noise to the VCO phase, we let the input phase be zero as shown in the simulation model where **phin** = 0.

This model, corresponding to the MATLAB m-file **pll1noise.m**[7], is a modified version of the MATLAB m-file **pll1sin.m** where equivalent noise (**nhat**) is added into the system. The equivalent noise, given in (6.8), is realized using the MATLAB expression

$$\textbf{nhat = -nd(i)*sin(phivco)+nq(i)*cos(phivco)} \tag{6.33}$$

where we assume that $A_c = 1$. The baseband noise components **nd** and **nq** are generated using the MATLAB function **randn()** with a specified value of input signal to noise ratio $(SNR)_i$.

---

[6]Again, the signals present at each node in the simulation model (phivco, s1, s2, ...) are identical to the corresponding variable names in the MATLAB simulation code.

[7]This MATLAB code is given at `http://www.morganclaypool.com/page/pll`.

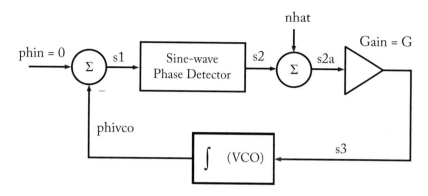

Figure 6.5: Simulation Model of First-order PLL with Additive Noise

6.3.2   SIMULATION RESULTS

The simulation for linear analysis is performed in a similar manner except that there is no sine-wave phase detector between **s1** and **s2** in the simulation model. Figures 6.6 and 6.7 show the variance of VCO phase and the noise bandwidth[8] B_L versus gain G for $(SNR)_i$ = -5 dB and $(SNR)_i$ = 5 dB, respectively.

We can see that, for fixed $(SNR)_i$, as gain G increases, both the VCO phase variance and noise bandwidth increase. From Figure 6.6, the values of VCO phase variance from both linear and nonlinear analysis are pretty close until the gain G is around 200 where the VCO phase variance is approximately 0.1. Similarly, the noise bandwidth from simulation follows the theoretical value until the gain G is around 100. From Figure 6.7, we see that when the value of VCO phase variance is very small, both linear and nonlinear analysis yield the same result. The noise bandwidth from simulation is very close to the theoretical value for all gain G.

Figures 6.8 and 6.9 show the histogram of phase error for $(SNR)_i$ = -5 dB and $(SNR)_i$ = 5 dB, respectively. The histogram is frequently used as an estimator of the probability density function (pdf) (18). The value of gain G used in the simulation is 100. The theoretical probability density function is plotted using (6.29) where the value[9] of ρ is calculated from the variance of phase error[10] obtained from the corresponding nonlinear simulation model. We can see that the results from a

---

[8]From Table 6.1, the theoretical value B_L = $G/4$. The value of B_L from simulation is calculated from

$$B_L = \frac{BW}{(SNR)_L/(SNR)_i} \tag{6.34}$$

where $BW = fs/2$, the sampling frequency fs used in the simulation model is 2000 Hz and $(SNR)_L = 1/(2\sigma_{\theta_n}^2)$ and $\sigma_{\theta_n}^2$ is the VCO phase variance.

[9]Recall that $\rho = 2(SNR)_L = 1/\sigma_{\theta_n}^2$.

[10]This variance is the same as the VCO phase variance $\sigma_{\theta_n}^2$ since the input phase is zero in the simulation model.

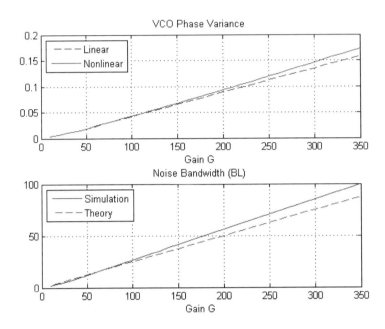

Figure 6.6: Variance of VCO Phase and Noise Bandwidth for $(SNR)_i$ = -5 dB

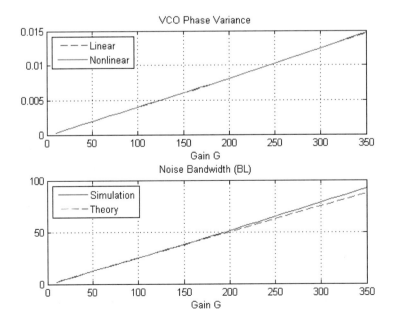

Figure 6.7: Variance of VCO Phase and Noise Bandwidth for $(SNR)_i$ = 5 dB

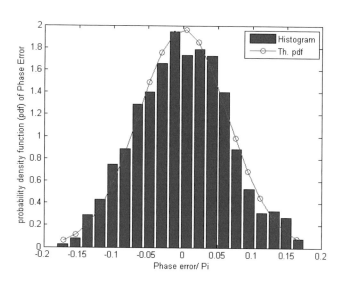

Figure 6.8: Histogram (an estimated pdf of ψ) for $(SNR)_i$ = -5 dB

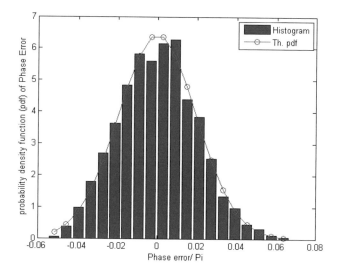

Figure 6.9: Histogram (an estimated pdf of ψ) for $(SNR)_i$ = 5 dB

nonlinear simulation model are comparable to the theoretical pdf derived from nonlinear analysis of the first-order PLL with additive noise.

6.4 PROBLEMS

6.1 Verify Equation (6.10).

6.2 Verify Equation (6.15).

6.3 Find the noise bandwidth B_L for the following:

Loop Type	f_n	ζ	λ	G	B_L
First order				1000	
Second order (Perfect)	$1000/(2\pi)$	0.2			
Second order (Perfect)	$1000/(2\pi)$	0.5			
Second order (Imperfect)	$1000/(2\pi)$	0.5	0.5		
Second order (Imperfect)	$1000/(2\pi)$	0.5	0.2		

6.4 Develop a MATLAB program to plot the **noise bandwidth** of the perfect second-order PLL and show that the damping value, ζ, of 0.5 yields the minimum noise bandwidth of $B_L/f_n = \pi$. Plot B_L/f_n vs. ζ.

6.5 Develop a MATLAB program to plot the **steady-state probability density function of** ψ (6.29) for values of $\rho = 0.1, 0.5, 1, 2$, and 4. Explain your plot for small and large values of ρ.

6.6 Develop a MATLAB program to plot **VCO phase variance** of the first-order PLL vs. **SNRi(dB)** for $-10 \leq$ SNRi(dB) ≤ 10 for both linear analysis and non-linear analysis. Explain your plot.

A P P E N D I X A

Complex Envelope and Phase Detector Models

The simulations considered in this book were based on the lowpass model, and, consequently, only lowpass signals were of interest. In the basic PLL model, only the loop input and the VCO output, which are inputs of the phase detector, are bandpass signals. Sampling these bandpass signals requires a very high sampling frequency in order to have the resulting sequence of samples accurately represent the bandpass signals. This is impractical in a simulation since a large number of samples must be generated, stored and processed for each second of the waveform. If the bandpass signals are represented by equivalent lowpass signals, the required sampling frequency becomes a function of only the modulation bandwidth. A significant time segment of the modulation signal can then be represented using a reasonable number of samples. The technique used to represent bandpass signals by equivalent lowpass signals lies in the concept of the complex envelope.

In this appendix[1], the general theory of the complex envelope is briefly reviewed. This is followed by a section treating the phase detectors used in the simulations presented in Chapter 4.

A.1   COMPLEX ENVELOPE

A general modulated bandpass signal can be represented by the expression

$$x(t) = A(t)\cos[2\pi f_c t + \phi(t)] \tag{A.1}$$

where $A(t)$ represents the amplitude modulation and $\phi(t)$ represents the phase deviation of the loop input, which may be the result of angle (frequency or phase) modulation, noise or a combination of modulation and noise. It is $\phi(t)$ that is the input to the lowpass PLL model. Using Euler's identity allows (A.1) to be written as

$$x(t) = \mathrm{Re}\left\{A(t)e^{j\phi(t)}e^{j2\pi f_c t}\right\} \tag{A.2}$$

---

[1]Much of the material in this chapter is based on Chapter 4 of

William H. Tranter, K. Sam Shanmugan, Theodore S. Rappaport, and Kurt L. Kosbar, *Principles of Communication Systems Simulation with Wireless Applications*, Prentice Hall, Professional Technical Reference, 2004, ISBN: 0-13-494790-8.

The student wishing to gain a deeper understanding of complex envelope representations of bandpass signals is referred to this material.

If a real bandpass signal is written in terms of the complex envelope[2], the form of the signal is

$$x(t) = \text{Re}\left\{\tilde{x}(t)e^{j2\pi f_c t}\right\} \tag{A.3}$$

where $\tilde{x}(t)$ is known as the complex envelope of the real signal $x(t)$. The complex envelope of the loop input corresponding to (A.1) is therefore

$$\tilde{x}(t) = A(t)e^{j\phi(t)} \tag{A.4}$$

which can easily be seen by simply comparing (A.2) and (A.3).

Writing the complex envelope signal in terms of its real and imaginary components yields

$$\tilde{x}(t) = x_d(t) + jx_q(t) \tag{A.5}$$

where the real part $x_d(t)$ is referred to as the direct component of the bandpass signal and the imaginary part $x_q(t)$ is referred to as the quadrature component of the bandpass signal. Substitution of (A.5) into (A.3) yields

$$x(t) = \text{Re}\left\{[x_d(t) + jx_q(t)][\cos 2\pi f_c t + j\sin 2\pi f_c t]\right\} \tag{A.6}$$

which gives

$$x(t) = x_d(t)\cos 2\pi f_c t - x_q(t)\sin 2\pi f_c t \tag{A.7}$$

This form is convenient for many applications.

Equation (A.3), or equivalently (A.6) and (A.7), show the manner in which the bandpass signal can be generated from the complex envelope $\tilde{x}(t)$. Note that knowledge of the carrier frequency is also required for complete determination of the bandpass signal. The direct and quadrature components can be obtained from the bandpass signal $x(t)$ by multiplication and lowpass filtering. Equation (A.3) can be written as

$$2x(t) = \tilde{x}(t)e^{j2\pi f_c t} + \tilde{x}^*(t)e^{-j2\pi f_c t} \tag{A.8}$$

Multiplication by $e^{-j2\pi f_c t}$ yields

$$2x(t)e^{-j2\pi f_c t} = \tilde{x}(t) + \tilde{x}^*(t)e^{-j4\pi f_c t} \tag{A.9}$$

Lowpass filtering will remove the last term, which is at twice the carrier frequency. Thus

$$\tilde{x}(t) = \text{Lp}\left\{2x(t)e^{-j2\pi f_c t}\right\} = x_d(t) + jx_q(t) \tag{A.10}$$

---

[2]Most introductory books on communication theory present basic treatments of the complex envelope such as (16), (21), (22)

where Lp {...} denotes the lowpass portion of the argument. Taking real and imaginary parts yields the direct and quadrature components of the complex envelope, respectively, as shown in Figure A.1(a). Equation (A.10) can also be written

$$x_d(t) + jx_q(t) = \text{Lp}\{2x(t)\cos2\pi f_c t\} - j\text{Lp}\{2x(t)\sin2\pi f_c t\} \tag{A.11}$$

Equating real and imaginary parts shows that

$$x_d(t) = \text{Lp}\{2x(t)\cos2\pi f_c t\} \tag{A.12}$$

and

$$x_q(t) = -\text{Lp}\{2x(t)\sin2\pi f_c t\} \tag{A.13}$$

Derivation of the direct and quadrature components of the complex envelope using (A.12) and (A.13) is illustrated in Figure A.1(b).

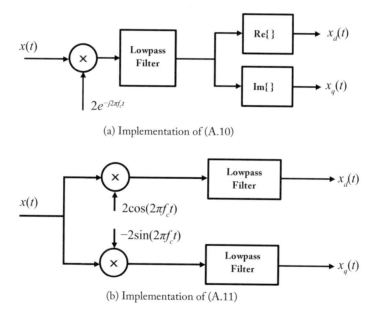

(a) Implementation of (A.10)

(b) Implementation of (A.11)

Figure A.1: Derivation of the Complex Envelope Components of a Bandpass Signal

A.2   PHASE DETECTOR REALIZATIONS

We now turn our attention to the specific problem of representing the PLL input and VCO output signals in a simulation. We assume that the loop input is represented by

$$x_{in}(t) = A(t)\cos[2\pi f_c t + \theta(t)] \tag{A.14}$$

where $A(t)$ represents the real envelope of the loop input signal. As was done in the previous section, we express the loop input in terms of the complex envelope. This gives

$$x_{in}(t) = \text{Re}\left\{ A(t)e^{j\phi(t)}e^{j2\pi f_c t} \right\} \tag{A.15}$$

or

$$x_{in}(t) = \text{Re}\left\{ \tilde{x}(t)e^{j2\pi f_c t} \right\} \tag{A.16}$$

where

$$\tilde{x}_{in}(t) = A(t)e^{j\phi(t)} \tag{A.17}$$

is the complex envelope of the loop input signal. In similar manner, the VCO output signal is

$$x_{vco}(t) = -A_v \sin[2\pi f_c t + \theta(t)] = \text{Im}\left\{ \tilde{x}_{vco}(t)e^{-j2\pi f_c t} \right\} \tag{A.18}$$

where

$$\tilde{x}_{vco}(t) = A_v e^{-j\theta(t)} \tag{A.19}$$

is the complex envelope of the VCO output. Multiplying (A.17) and (A.19) yields

$$\tilde{x}_{in}(t)\tilde{x}_{vco}(t) = A(t)A_v e^{j[\phi(t)-\theta(t)]} \tag{A.20}$$

Taking the imaginary part gives

$$\text{Im}\left\{\tilde{x}_{in}(t)\tilde{x}_{vco}(t)\right\} = A(t)A_v \sin[\phi(t) - \theta(t)] \tag{A.21}$$

which is identical to the phase detector output as expressed by (2.3) except for the assumed time-variation of the real envelope of the PLL input signal.

There are two important cases to consider; the case in which $A(t)$ may be treated as a constant and the case in which $A(t)$ is time-varying. If $A(t)$ is a constant, the phase detector output is simply

$$e_d(t) = A_c A_v \sin[\phi(t) - \theta(t)] \tag{A.22}$$

as expressed in (2.3). In this case, the multiplicative constant $A_c A_v$ can be absorbed into the loop gain as shown in (2.7). The baseband model illustrated in Figure A.2 results.

Next, consider the case in which both the amplitude and the phase deviation of the input are time varying, and both must be accurately modeled. In this case, the loop input must consist of two components, $A(t)$ and $\phi(t)$. This is easily accomplished using a complex envelope signal representation. In the baseband (lowpass signal) PLL model, the loop input is the complex envelope

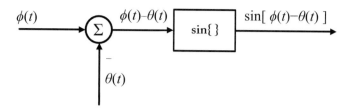

Figure A.2: Baseband Phase Detector Model for Constant-Amplitude Signals

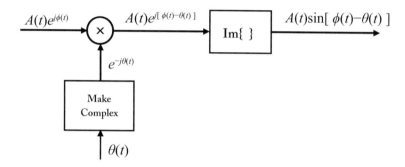

Figure A.3: Baseband Phase Detector Model for Time-Varying Amplitude PLL Input Signals

of the actual bandpass signal. The amplitude of this complex signal is $A(t)$ and the phase is $\phi(t)$. Letting $A_v = 1$ in (A.20) gives

$$\tilde{x}_{in}(t)\tilde{x}_{vco}(t) = A(t)e^{j[\phi(t)-\theta(t)]} \tag{A.23}$$

where, as previously noted, we can absorb the VCO amplitude A_v into the loop gain as noted in (2.7). Taking the imaginary part yields the phase detector output for the time-varying PLL input amplitude case. The result is

$$\mathrm{Im}\{\tilde{x}_{in}(t)\tilde{x}_{vco}(t)\} = A(t)\sin[\phi(t)-\theta(t)] \tag{A.24}$$

since $A(t)$ is real. This yields the phase detector model for the time-varying amplitude signals shown in Figure A.3. This model is used in the Costas BPSK demodulator presented in Chapter 5. The required complex envelope of the loop input signal is generated in the simulation preprocessor.

APPENDIX B

Loop Filter Implementations

In this appendix[1] the loop filter implementations are developed[2]. Trapezoidal integration is used in the numerical simulations, and is derived in Section B.1. The various loop filters used in the second-order and third-order PLL simulations are then developed in later sections. Many discrete-time approximations exist for continuous-time integration. For a discussion of these techniques from a simulation perspective, see Tranter, et.al (18).

B.1 TRAPEZOIDAL INTEGRATION

Integration is used in both the PLL filter models and in the model for the loop VCO. In all cases, trapezoidal integration is used as the numerical approximation to the required continuous-time integration process. Trapezoidal integration is equivalent to the bilinear z-transform model for the integration process, and, therefore, aliasing errors are eliminated. In addition there is no phase distortion induced by trapezoidal integration. These desirable characteristics are achieved, however, at the cost of distortion of the ideal amplitude response resulting from the frequency warping of the bilinear z-transform. If the sampling frequency is sufficiently high, relative to the signal bandwidth, the effect of this frequency warping is negligible (17).

As shown in Figure B.1, the value of the integrator output at sample index n is given by

$$y[n] = y[n-1] + \Delta(n-1, n) \tag{B.1}$$

where $\Delta(n-1, n)$ is the approximation to the value of the integral in the range $\{(n-1)T, nT\}$ where T is the sampling period. The discrete-time approximation to

$$\Delta(n-1, n) \approx \int_{(n-1)T}^{nT} x(t)dt \tag{B.2}$$

---

[1]Much of the material in this chapter is based on Chapter 5 of

William H. Tranter, K. Sam Shanmugan, Theodore S. Rappaport, and Kurt L. Kosbar, *Principles of Communication Systems Simulation with Wireless Applications*, Prentice Hall, Professional Technical Reference, 2004, ISBN: 0-13-494790-8.

The student wishing to gain a deeper understanding of synthesis techniques of discrete-time filters is referred to this material.

[2]Almost any book containing materials on digital signal processing and digital filters will have material relevant to this chapter. Examples of such books include (17), (20), (23), and (24).

is given by

$$\Delta(n-1, n) = \frac{T}{2}(x[n-1] + x[n]) \tag{B.3}$$

and is represented by the shaded area in Figure B.1.

Substitution of (B.3) into (B.1) yields

$$y[n] - y[n-1] = \frac{T}{2}(x[n-1] + x[n]) \tag{B.4}$$

Taking the Z-transform gives

$$(1 - z^{-1})Y(z) = \frac{T}{2}(1 + z^{-1})X(z) \tag{B.5}$$

so that

$$H(z) = \frac{Y(z)}{X(z)} = \frac{T}{2}\frac{1 + z^{-1}}{1 - z^{-1}} \tag{B.6}$$

is the transfer function of the trapezoidal integrator. Note that this is the bilinear Z-transform of $1/s$ where s is the Laplace frequency domain variable.

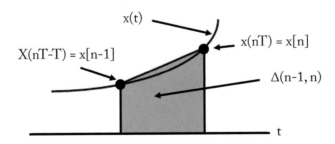

Figure B.1: Illustration of Trapezoidal Integration

A computationally efficient algorithm for implementing the trapezoidal integrator can be derived from the transformed Direct Form II signal flow graph for the integrator (20). The signal flow graph of the transformed Direct Form II is generated by starting with the signal flow graph of the Direct Form II implementation, which is illustrated in Figure B.2(a). The transformed Direct Form II structure is derived from the Direct Form II structure by reversing the direction of the signal flow in each branch of the signal flow graph and interchanging the input and output. These operations yield the signal flow graph illustrated in Figure B.2(b). Redrawing the signal flow graph

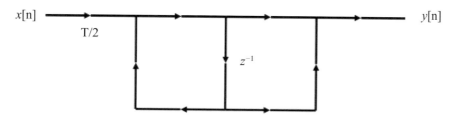

(a) Direct Form II Signal Flow Graph for Trapezoidal Integrator

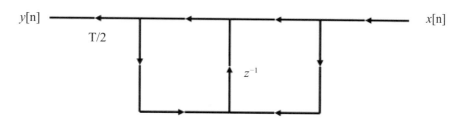

(b) Transposed Direct Form II Signal Flow Graph for Trapezoidal Integrator

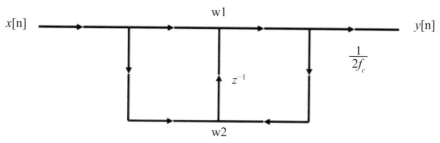

(c) Transposed Form II Signal Flow Graph for a Trapezoidal Integrator in Standard Form

Figure B.2: Derivation of the Transposed Direct Form II Signal Flow Graph for a Trapezoidal Integrator

so that the basic direction of signal flow is from left to right yields the transformed Direct Form II signal flow graph as shown in Figure B.2(c).

Note that in Figure B.2(c) two intermediate nodes, denoted **w1** and **w2**, are shown and that the sampling period has been replaced by the reciprocal of the sampling frequency. It follows from the signal flow graph illustrated in Figure B.2(c) that

$$w1 = x[n] + w2 \qquad (B.7)$$

and

$$w2 = x[n] + w1 \tag{B.8}$$

The integrator output is given by

$$y[n] = w1/(2f_s) \tag{B.9}$$

where f_s denotes the sampling frequency. Thus, denoting the input sample by x and the output sample by y yields the following code structure:

```
          ⋮
w2 = 0
for i=1:npts

               ⋮
          w1   =   x(i)+w2
          w2   =   x(i)+w1
          y(i) =   w1/(2*fs)
               ⋮

end
```

Note that **w2** must be initialized prior to the first execution, $i = 1$, of the simulation loop. Note also that **w2** is computed after **w1** but that **w2** is not used until the next execution of the simulation loop. Thus, for $i = n$, (B.7) is

$$w1[n] = x[n] + w2[n-1] \tag{B.10}$$

so that the unit delay z^{-1} is implemented through the order in which **w1** and **w2** are computed.

B.2   THE LOOP FILTER FOR THE PERFECT SECOND-ORDER PHASE-LOCKED LOOP

The required loop filter for the perfect second-order PLL is defined by

$$F(s) = 1 + \frac{a}{s} \tag{B.11}$$

Recognizing that division by s is equivalent to integration shows that the loop filter can be implemented as shown in Figure B.3. In Figure B.3, as well as those which follow, **gi** is used to denote the ith internal node and **x** and **y** are used to denote the filter input and output nodes, respectively.

Using the trapezoidal integration algorithm developed in the previous section, the algorithm for implementing the loop filter is easily developed. The result is

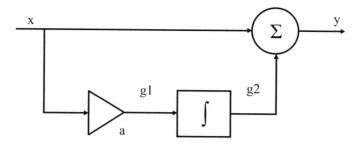

Figure B.3: Loop Filter for Perfect Second-Order Loop

$$
\begin{aligned}
g1 &= a^*x \\[1em]
w1 &= g1+w2 \\
w2 &= g1+w1 \\
g2 &= w1/(2^*fs) \\[1em]
y &= x+g2
\end{aligned}
$$

in which **fs** is the sampling frequency and where the lines of code used to implement the integrator have been grouped together. As discussed previously, **w2** must be initialized prior to executing the simulation loop.

B.3   THE LOOP FILTER FOR THE IMPERFECT SECOND-ORDER PHASE-LOCKED LOOP

The required loop filter for the imperfect second-order PLL is defined by

$$
F(s) = \frac{s+a}{s+\lambda a} \tag{B.12}
$$

where λ is the fractional offset of the pole relative to the location of the zero. As mentioned in Chapter 2, we are typically interested in small values of λ.

The implementation of the loop filter used in the imperfect second-order PLL simulations given in Chapter 4 is obtained by using long division to express the transfer function in the form

$$
F(s) = 1 + (1-\lambda)a \frac{1}{s+\lambda a} \tag{B.13}
$$

or

$$
F(s) = 1 + (1-\lambda)a F_1(s) \tag{B.14}
$$

where

$$F_1(s) = \frac{1}{s + \lambda a} = \frac{Y_1(s)}{X_1(s)} \tag{B.15}$$

The preceding expression can be written as

$$Y_1(s)(s + \lambda a) = X_1(s) \tag{B.16}$$

or

$$Y_1(s) = \frac{1}{s}[X_1(s) - \lambda a Y_1(s)] \tag{B.17}$$

This yields the differential equation

$$y_1(t) = \int [x_1(t) - \lambda a y_1(t)]dt \tag{B.18}$$

which can be implemented as shown in Figure B.4. Combining (B.14) and (B.15) yields the implementation of the loop filter for the imperfect second-order PLL as shown in Figure B.5. The algorithm for implementing the filter is easily derived. The result is

$$
\begin{aligned}
\text{g1} &= (1 - \lambda)^* a^* x \\
\text{g2} &= -\lambda^* a^* \text{g4} \\
\text{g3} &= \text{g1+g2} \\
\\
\text{w1} &= \text{g3+w2} \\
\text{w2} &= \text{g3+w1} \\
\text{g4} &= \text{w1/(2*fs)} \\
\\
\text{y} &= \text{x+g4}
\end{aligned}
$$

It is clear that both **g4** and **w2** must be initialized prior to the initial execution of the simulation loop. Note that as $\lambda \to 0$ the imperfect loop filter converges to the perfect loop filter.

B.4   THE PERFECT THIRD-ORDER LOOP FILTER

As shown in Chapter 2, the perfect third-order PLL requires a loop filter of the form

$$F(s) = 1 + \frac{a}{s} + \frac{b}{s^2} = 1 + \frac{1}{s}\left(a + \frac{b}{s}\right) \tag{B.19}$$

Once again, by realizing that division by s is equivalent to integration, we can easily develop the implementation of the filter. The result is shown in Figure B.6. The algorithm for implementing the filter is

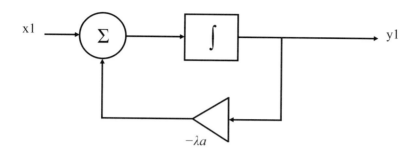

Figure B.4: Implementation of Equation (B.18)

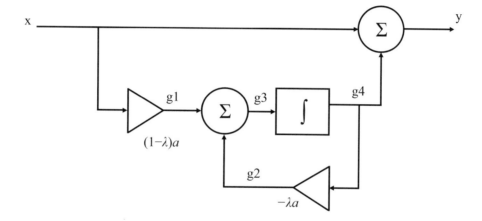

Figure B.5: Implementation of the Loop Filter for the Imperfect Second-Order PLL

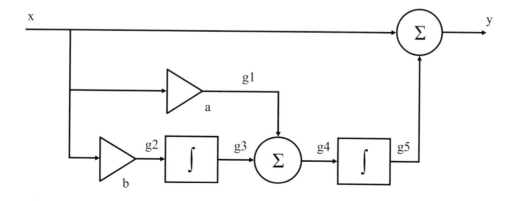

Figure B.6: Loop Filter for Third-Order PLL

$$\begin{aligned}
g1 &= a^*x \\
g2 &= b^*x
\end{aligned}$$

$$\begin{aligned}
w1a &= g2+w2a \\
w2a &= g2+w1a \\
g3 &= w1a/(2^*fs)
\end{aligned}$$

$$g4 = g1+g3$$

$$\begin{aligned}
w1b &= g4+w2b \\
w2b &= g4+w1b \\
g5 &= w1b/(2^*fs)
\end{aligned}$$

$$y = x+g5$$

It is clear that **w2a** and **w2b** must be initialized prior to the initial execution of the simulation loop.

APPENDIX C

SIMULINK Examples

We consider several SIMULINK realizations of systems previously considered in Chapter 2. For those unfamiliar with SIMULINK, the example simulations which follow show the utility of SIMULINK and provide simulation templates that could serve as building blocks for more complex systems. Three SIMULINK simulations are included in this appendix and are defined in the following SIMULINK files[1]:

simpll2.mdl - SIMULINK for perfect second-order PLL
simpll2tdel.mdl - SIMULINK for perfect second-order PLL with transport delay
simpll3.mdl - SIMULINK for perfect third-order PLL

C.1   THE PERFECT SECOND-ORDER PLL

The SIMULINK realization of the perfect second-order PLL is illustrated in Figure C.1. The *XY Graph* block is configured to plot the phase-plane for the system, which is defined as the frequency error versus the phase error. The X input, the phase error, is taken directly from the output of the phase detector as shown. The Y input, the frequency error, is generated as shown. It should be noted that the frequency error is expressed in Hertz.

The block labeled *floating scope* is useful for quickly observing waveforms at various points in the system as the simulation progresses. One should refer to the SIMULINK manual for details on its use.

The blocks for the loop gain, G, and the filter constant, a, are "masked" since it is more convenient to enter the loop natural frequency f_n and the loop damping factor ζ than to enter the values for G and a. The SIMULINK user *clicks* on the appropriate blocks and enters the desired values of f_n and ζ. G and a are then calculated according to the expressions

$$G = 4\pi \zeta f_n \tag{C.1}$$

and

$$a = \frac{4\pi^2 f_n^2}{G} = \frac{\pi f_n}{\zeta} \tag{C.2}$$

---

[1]These SIMULINK files can be found at http://www.morganclaypool.com/page/pll.

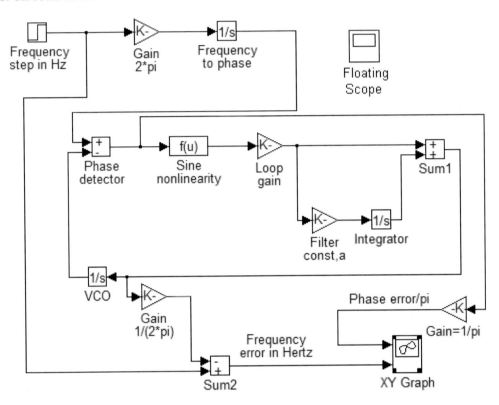

Figure C.1: SIMULINK Block Diagram for Perfect Second-Order PLL

as were developed in Chapter 2. They are reproduced here for convenience.

The initial step in executing a SIMULINK simulation is to select the simulation parameters. The start time and the stop time are straightforward to select. The integration methods supplied with SIMULINK can be either variable-step or fixed-step methods. The default is a variable step method. This leads, of course, to quite different simulations than the MATLAB simulations developed previously since the MATLAB simulations were based on the simple trapezoidal integrator, which is a fixed-step algorithm. In SIMULINK, the user must select a minimum and a maximum step size, a tolerance and an integration method. The step size should be related to the bandwidth of the signals expected in the system in order to avoid aliasing errors. In variable step integration methods, the step size is continually adjusted between the minimum and the maximum specified values such that the relative error (tolerance) is kept within the bound specified. The student should refer to Integrator in the SIMULINK manual.

The simulation **simpll2.mdl** was executed with the following parameters:

Start Time:	0.0
Stop Time:	1.0
Min Step Time:	0.0001
Max Step Size:	0.001
Tolerance:	1e-3
Integration (Solver):	ode45
	(based on an explicit Runge-Kutta (4,5) formula)

The result is the phase plane plot, which appears in the X-Y Graph SIMULINK block, and is shown in Figure C.2 where the vertical axis is the frequency error in Hertz and the horizontal axis is the phase error in radians normalized by π. Since the final value of X, as expressed in the X-Y Graph box is 6, the steady-state phase error is 6π. Since each cycle slipped adds 2π to the steady-state error, it follows that three cycles are slipped in the acquisition process.

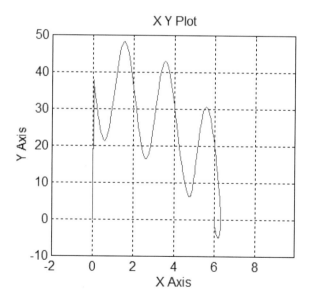

Figure C.2: Phase Plane Plot Developed Using the SIMULINK Program of Figure C.1

Clearly, one should at this point compare Figure C.2 with Figure C.3, the MATLAB result obtained using the fixed-step trapezoidal integration method. While the two results are in basic agreement, and each result illustrates that three cycles are slipped in the acquisition process, there are subtle differences. Consider, for example, the maximum and the minimum frequency error within each of

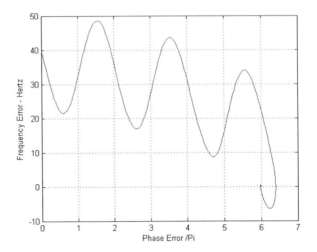

Figure C.3: Extended Phase Plane Plot using the Simulation Model of Figure 4.2 when $\lambda = 0$

the cycles slipped. One observes approximate agreement but not exact agreement. One important source of these differences lies in the differing characteristics of the integration algorithms. The student should explore the impact of choosing different integration algorithms and the impact of using various values for maximum and minimum step sizes and tolerance.

C.2   THE PERFECT SECOND-ORDER PLL WITH TRANSPORT DELAY

We now modify the previous model to include transport delay. Transport delay is, fortunately, a built-in SIMULINK model so the modification is indeed simple. The transport delay is set equal to 0.005 seconds to correspond with the example presented in Section 2.3.5[2]. The simulation diagram is shown in Figure C.4.

Observing Figure C.4 shows an additional modification of the previous simulation. A block labeled *out5* has been added to the simulation diagram. This block results in the samples generated by the simulation at the point to which the block is attached to be written to the MATLAB workspace. Since the block is connected to the VCO input (after passing through the gain of $1/2\pi$), the data written to the MATLAB workspace is the VCO frequency deviation in Hertz.

---

[2]Recall that for the example executed in Section 2.3.5 the sampling frequency was 2000 samples per second. Thus the fixed step size was 0.5 milliseconds. The transport delay was 9 sample periods, which gives a transport delay of 4.5 milliseconds. Recall, however, that there is an *implicit* delay of 1 sample period around the loop given the technique used to develop the simulation program. Taking into account this additional delay of one sample period yields a total transport delay of 5 milliseconds.

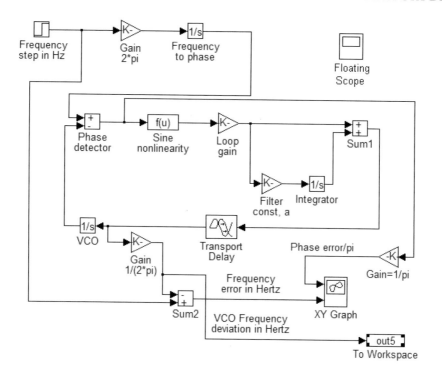

Figure C.4: SIMULINK Block Diagram for Perfect Second-Order PLL with Transport Delay

The simulation illustrated in Figure C.4 is given in **simpll2tdel.mdl**. It was executed with the following parameters:

Start Time:	0.0
Stop Time:	1.0
Min Step Time:	0.0001
Max Step Size:	0.001
Tolerance:	1e-3
Integration (Solver):	ode45
	(based on an explicit Runge-Kutta (4,5) formula)

The resulting phase plane is shown in Figure C.5. In comparing this result to the fixed-step trapezoidal integration result developed in Section 2.3.5 and illustrated in Figure C.6, we again see basic agreement. In both cases, 9 cycles are slipped in the acquisition process.

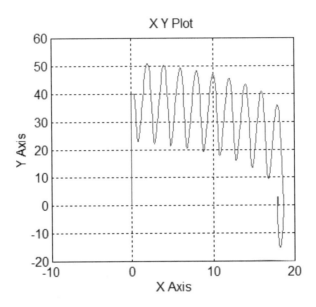

Figure C.5: Phase-Plane Plot Developed Using the SIMULINK Program of Figure C.4

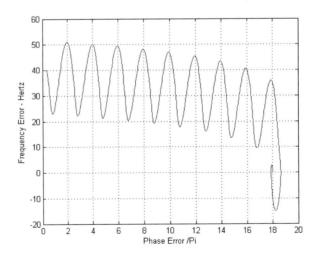

Figure C.6: Extended Phase Plane Plot using the Simulation Model of Figure 4.4 when $\lambda = 0$

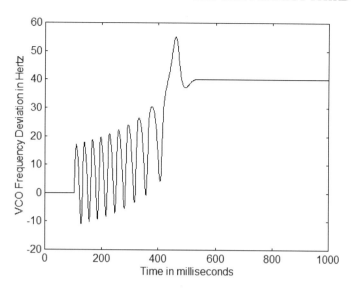

Figure C.7: VCO Frequency Deviation obtained by the SIMULINK Program of Figure C.4

Using the data returned to the MATLAB workspace, we can easily plot the VCO frequency deviation using the command

```
plot(out5)
xlabel('Time in milliseconds')
ylabel('VCO Frequency Deviation in Hertz')
```

The result is illustrated in Figure C.7. From Figure C.7, and recalling that the frequency step occurs 10 percent into the simulation which is at the 10 millisecond point in this case, we see that it takes approximately 450 milliseconds for the loop to acquire the signal after the frequency step is applied. By executing the simulation for different values of transport delay, one can immediately see the impact of various transport delays. Although somewhat dangerous, we can compare Figure C.7 to Figure C.8 and see the additional acquisition time resulting from the transport delay[3].

C.3   THE PERFECT THIRD-ORDER PLL

We now turn our attention to the perfect third-order PLL. While we could implement the loop filter as a ratio of polynomials in s using the SIMULINK block ***Transfer Function***, we choose

---

[3]Such comparisons are dangerous because the result shown in Figure C.8 was developed for fixed-step trapezoidal integration while the result shown in Figure C.7 was developed using variable-step fifth-order Runge-Kutta integration. If the sampling frequencies are sufficiently high, the differences should be negligible, but one should be very careful comparing results of simulations using such different integration algorithms.

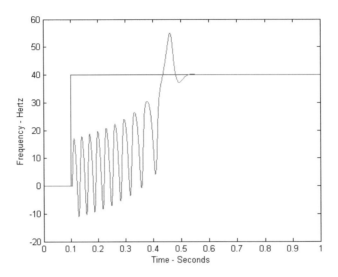

Figure C.8: VCO Frequency Deviation using the Simulation Model of Figure 4.4 when $\lambda = 0$

to implement the loop filter using discrete integrators so that it matches up more closely with the previously considered MATLAB simulations. The result is shown in Figure C.9. The same parameters (Loop Gain and the Filter Parameters a and b) used to obtain Figure 2.14 were applied to the SIMULINK program of Figure C.9. Executing the simulation with the same parameters used in the previous simulations, namely,

Start Time:	0.0
Stop Time:	1.0
Min Step Time:	0.0001
Max Step Size:	0.001
Tolerance:	1e-3
Integration (Solver):	ode45
	(based on an explicit Runge-Kutta (4,5) formula)

results in the simulation result shown in Figure C.10. Comparing Figure C.10 to the MATLAB result, Figure 2.14, obtained with fixed-step trapezoidal integration, shows basic agreement although, of course, subtle differences can be seen. As mentioned previously, these differences are most likely due to the different integration algorithms.

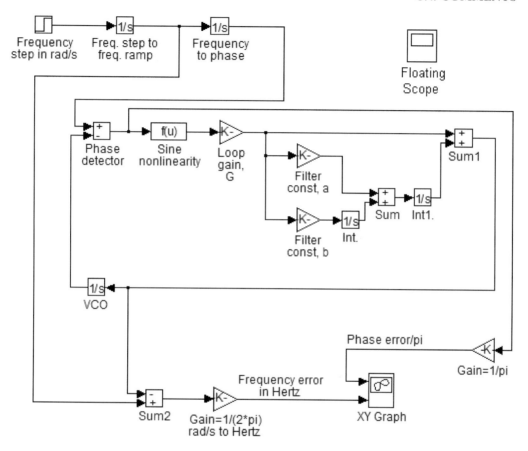

Figure C.9: SIMULINK Block Diagram for Perfect Third-Order PLL

C.4 COMMENTS

SIMULINK is certainly a valuable tool for quickly investigating the dynamic behavior of a complex system. The process of developing simulations by manipulating system building blocks on a palate is appealing. Note that all three operations of preprocessing, simulation and postprocessing are present in SIMULINK although they are merged in the sense that only a single program, rather than three separate programs, are executed. It should be pointed out that a separate MATLAB postprocessor can be used with SIMULINK by returning the appropriate vectors to the MATLAB workspace as was done in Section C.2. As a matter of fact if all necessary vectors are returned to the MATLAB workspace, the external postprocessor **pllpost.m** could be used with any of the three simulations presented in this appendix.

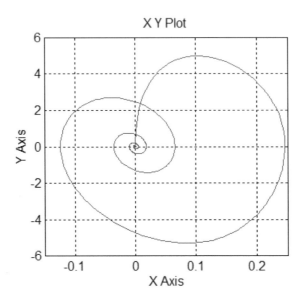

Figure C.10: Phase-Plane Plot Corresponding to the SIMULINK Program of Figure C.9

It is suggested that the interested student experiment with the different integration algorithms native to SIMULINK and with different sets of integration parameters. A combination of the previously developed MATLAB simulations and the SIMULINK simulations presented here allows one to compare fixed-step and variable-step integration algorithms. The approach taken in the SIMULINK simulations presented in this appendix is that of modeling a **continuous-time system**. While the physical PLLs are most likely continuous-time systems, one could consider a discrete-time model and then simulate the discrete-time model using SIMULINK. If one takes this approach, the integration algorithms could be represented by the appropriate z-domain transfer function in SIMULINK, and other comparisons could be made. As an example, one could use trapezoidal or even rectangular integration.

An important aspect of simulation is validation, which is the process of ensuring that the simulation results are reasonable and consistent with known theory. One way of identifying certain types of errors is to perform multiple simulations using different tools. In that sense, the consistency between the MATLAB results and the SIMULINK results provides partial validation.

From time-to-time in this appendix, we have compared SIMULINK results to the MATLAB results presented elsewhere in this book. Subtle differences were noted. The question *Which results are correct?* naturally arises. One could correctly say that none are correct in the true sense of the word, and yet all are correct. None of the results presented in this book are correct since none of the simulation programs exactly models the physical device in complete detail. All of the results presented here are correct in the sense the results are consistent with the simulation model, including all of the approximations made in the development of that model. All of the MATLAB and SIMULINK results illustrate the general behavior of the tracking systems with the sets of inputs considered. There will always be a difference between the physical device and the simulation model. ***How great can this difference be and still have useful results from the simulation model?*** The answer to this question depends upon the application since different applications can tolerate different levels of error. There are many trade offs involving model complexity, simulation runtime and error levels. To properly evaluate all of the trade offs is the art of engineering.

APPENDIX D

MATLAB and SIMULINK Files

All simulations can be found at `http://www.morganclaypool.com/page/pll`. There are two folders: the first folder, **PLLloops**, contains the code for the MATLAB simulations and the second folder, **PLLsims**, contains the SIMULINK simulations. The contents of these folders are as follows:

D.1   MATLAB FILES

Contents of **PLLloops**

pllpre.m	Preprocessor for first-order, second-order and third-order PLL simulations
pll1sin.m	Simulation code for first-order PLL with a sine-wave phase detector
pll2sin.m	Simulation code for second-order PLL with a sine-wave phase detector
pll2tri.m	Simulation code for second-order PLL with a triangular-wave phase detector
pll2saw.m	Simulation code for second-order PLL with a sawtooth-wave phase detector
pll2tdel.m	Simulation code for second-order PLL with a sine-wave phase detector and transport delay
pll3sin.m	Simulation code for third-order PLL with a sine-wave phase detector
pllpost.m	Postprocessor for PLL simulations
cpllpre.m	Preprocessor for Costas PLL simulation
cpll.m	Simulation code for Costas PLL
cpllpost.m	Postprocessor for Costas PLL simulation

Contents of **PLLloops** (*(cont.)*)

qcpllpre.m	Preprocessor for QPSK loop simulation
qcpll.m	Simulation code for QPSK loop
qcpllpst.m	Postprocessor for QPSK loop simulation
npll.m	Simulation code for N-Phase tracking loop (Note: The N-Phase tracking loop uses the preprocessor and postprocessor for the QPSK loop.)
ppplot.m	Support file for the PLL postprocessor (Primary routine for developing phase plane plots)
pplane.m	Support file for the PLL postprocessor (Routine for developing phase plane plots mod(2π))
qpsk.m	Support file for **qcpll.m** (Implements QPSK signal set)
randombinary.m	Support file for both the Costas PLL and the QPSK loop (Implements random modulation)

D.2 SIMULINK FILES

Contents of **PLLsims**

simpll2.mdl	SIMULINK simulation for second-order PLL
simpll2tdel.mdl	SIMULINK simulation for a perfect second-order PLL with transport delay
simpll3.mdl	SIMULINK simulation for third-order PLL

Bibliography

[1] Gardner, Floyd M., *Phaselock techniques*, 3rd edition, Wiley, John and Sons, 2005. DOI: 10.1002/0471732699 23, 46, 78, 82, 84, 85

[2] Gardner, Floyd M., *Phaselock techniques*, 2nd edition, Wiley, John and Sons, 1979.

[3] Blanchard, Alain, *Phase-locked loops : Application to Coherent Receiver Design*, Wiley, 1976. 23

[4] Best, Roland E., *Phase-Locked Loops: Design, Simulation, and Applications*, 6th edition, McGraw-Hill, 2007. 46

[5] Encinas, J.B., *Phase Locked Loops*, Chapman and Hall, 1993.

[6] Heinrich, Meyr and Ascheid, Gerd, *Synchronization in Digital Communications (Volume 1): Phase-, Frequency-Locked Loops, and Amplitude Control*, Wiley Interscience, 1990. 23, 82, 86

[7] Viterbi, A.J., *Principles of Coherent Communications*, McGraw-Hill, 1966. 26, 77, 78, 83

[8] Lindsey, W. C. and Simon, M. K., *Telecommunication Systems Engineering*, Prentice-Hall, 1973. 41, 82

[9] Bennedetto, S., Biglieri E. and Castellini, V., *Digital Transmission Theory*, Prentice-Hall, 1987. 36

[10] Braun, W. R. and Lindsey, W. C., *Carrier Synchronization Techniques for Unbalanced QPSK Signals–Part I*, IEEE Trans. on Communications, Vol. COM-26, No. 9, September 1978. DOI: 10.1109/TCOM.1978.1094235 36

[11] Egan, W.F., *Phase-Lock Basics* (Second Edition), Wiley-Interscience, 2008. 86

[12] Viterbi, A.J., *Phase-locked loop dynamics in the presence of noise by Fokker-Planck techniques*, Proc. IEEE vol. 51, pp. 1737-1753, Dec 1963. DOI: 10.1109/PROC.1963.2686 83, 84

[13] Stensby, J. L., *Phase-locked loops : theory and applications*, CRC Press, 1997. 82

[14] Barnes, J. A., et al., *Characterization of Frequency Stability*, IEEE Trans. Instrum. Meas. IM-20, 105-120, May 1971. 85

[15] Kroupa, V. F., ed., *Frequency Stability: Fundamentals and Measurement*, Reprint volume, IEEE Press, New York, 1983. 84

[16] Ziemer, R. E. and Tranter, W. H., *Principles of Communications: Systems, Modulation and Noise*, 4th Edition, Wiley, 1995, pp. 651-654. 1, 92

[17] Ziemer, R. E., Tranter, W. H., and Fannin, D. R., *Signals and Systems: Continuous and Discrete*, 3rd Edition, Prentice-Hall, 1993. 14, 97

[18] Tranter, W. H., Shanmugan K. S., Rappaport T. S., and Kosbar, K. L., *Principles of Communication Systems Simulation: with Wireless Applications*, Prentice-Hall, 2004. 28, 56, 59, 87, 97

[19] Ziemer, R. E. and Peterson, R.L., *Introduction to Digital Communications*, 2nd Edition, Prentice-Hall, 2001. 29

[20] Oppenheim, A. V. and Schafer, R. W., *Discrete-Time Signal Processing*, Prentice Hall, 1989. 97, 98

[21] Haykin, S., *Communication Systems* (Third Edition), Wiley, 1994. 92

[22] Jeruchim, M. C., Balaban, P. and Shanmugan, K. S., *Simulation of Communication Systems*, Plenum Press, 1992. 92

[23] Proakis, J. G. and Manolakis, D. G., *Digital Signal Processing: Principles, Algorithms and Applications* (Second Edition), Macmillan, 1992. 97

[24] Strum, R. D. and Kirk D. E., *First Principles of Discrete Systems and Digital Signal Processing*, Addison-Wesley, 1988. 97

Authors' Biographies

WILLIAM H. TRANTER

William H. (Bill) Tranter received the Ph.D. degree in 1970. He joined the faculty of the University of Missouri-Rolla in 1969. From 1980 to 1985, he served as Associate Dean of Engineering with responsibility for research and graduate affairs. He was appointed Schlumberger Professor of Electrical Engineering in 1985 and served in that position until his early retirement from UMR in 1997.

In 1996-7, Bill served as an Erskine Fellow at Canterbury University in Christchurch, New Zealand. In 1997, he joined the Electrical Engineering faculty of the Virginia Polytechnic Institute and State University, (Virginia Tech), in Blacksburg, VA, as the Bradley Professor of Communications. In 2009, Bill took an IPA leave from Virginia Tech and now serves as Program Director for Communications, Information Theory, and Coding at the National Science Foundation.

His research interests are digital signal processing and computer-aided design of communication systems applied to wireless communications systems. He has authored numerous technical papers and is the co-author of three textbooks: *Principles of Communications: Systems, Modulation and Noise* (Wiley, 2002), *Signals and Systems* (Prentice-Hall, 1998), and *Simulation of Communication Systems with Applications to Wireless Communications* (Prentice-Hall).

He has held many positions within the IEEE Communications Society including Director of Journals, Director of Education, and as a member and chair of a number of technical committees. He served as a member of the Board of Governors of the IEEE Communications Society, and as Vice President—Technical Activities. For eleven years, he served as Editor-in-Chief of the IEEE JOURNAL ON SELECTED AREAS IN COMMUNICATIONS. In that position, he founded the IEEE TRANSACTIONS ON WIRELESS COMMUNICATIONS. He recently completed a three-year term as a member of the IEEE FELLOW Committee for the IEEE Board of Directors.

He was named a Fellow of the IEEE in 1985 and has received numerous awards, including the James McLellan Meritorious Service Award, the IEEE Exemplary Publications Award, the IEEE Centennial Medal, and the IEEE Third Millennium Medal.

RATCHANEEKORN THAMVICHAI

Ratchaneekorn Thamvichai received her Ph.D. degree in electrical engineering from University of Colorado, Boulder in 2002. She received her M.S. degree and BEng degree in electrical engineering from Stanford University and Chulalongkorn University, respectively. Currently, she is Associate

Professor in the Electrical and Computer Engineering department at Saint Cloud State University in Minnesota.

From 2009 to July 2010, she was a Visiting Research Associate Professor in the Wireless@VT group in the Bradley Department of Electrical and Computer Engineering at Virginia Tech where she had an opportunity to work with Dr. Bose and Dr. Tranter.

Her research interests include signal classification and signal processing for cognitive radios and one-dimensional and two-dimensional digital signal processing.

TAMAL BOSE

Tamal Bose received the Ph.D. degree in electrical engineering from Southern Illinois University in 1988. After a faculty position at the University of Colorado, he joined Utah State University in 2000, where he served as the Department Head and Professor of Electrical and Computer Engineering from 2003-2007. Currently, he is Professor in the Bradley Department of Electrical and Computer Engineering at Virginia Tech. He is the Associate Director of Wireless@VT and Director of the NSF center site WICAT@VT.

The research interests of Dr. Bose include signal classification for cognitive radios, channel equalization, adaptive filtering algorithms, and nonlinear effects in digital filters. He is author of the text ***Digital Signal and Image Processing***, John Wiley, 2004. Dr. Bose served as the Associate Editor for the IEEE Transactions on Signal Processing from 1992 to 1996. He is currently on the editorial board of the IEICE Transactions on Fundamentals of Electronics, Communications and Computer Sciences (Japan) and the Journal of Electrical and Computer Engineering. He also served on the organizing committees of several international conferences and workshops.

Printed in the United States
by Baker & Taylor Publisher Services